Agricultural Applications in Green Chemistry

ACS SYMPOSIUM SERIES **887**

Agricultural Applications in Green Chemistry

William M. Nelson, Editor
Waste Management and Research Center

Sponsored by the
ACS Division of Industrial and Engineering
Chemistry, Inc.

American Chemical Society, Washington, DC

Library of Congress Cataloging-in-Publication Data

Agricultural applications in green chemistry / William M. Nelson, editor ; sponsored by the ACS Division of Industrial and Engineering Chemistry, Inc.

    p. cm.—(ACS symposium series ; 887)

    "Developed from a symposium sponsored by the Division of Industrial and Engineering Chemistry, Inc. at the 223rd National Meeting of the American Computer Society, Orlando, Florida, April 7–11, 2002"—T.p. verso.

    Includes bibliographical references and index.

    ISBN 0–8412–3828–6

    1. Agricultural chemistry—Industrial applications—Congresses. 2. Environmental chemistry—Industrial applications—Congresses.

    I. Nelson, William M. II. American Chemical Society. Division of Industrial and Engineering Chemistry, Inc. III. American Chemical Society. Meeting (223rd : 2002 : Orlando, Fla.) IV. Series.

S583.2.A375   2004
630′.2′4—dc22   2004046176

The paper used in this publication meets the minimum requirements of American National Standard for Information Sciences—Permanence of Paper for Printed Library Materials, ANSI Z39.48–1984.

Copyright © 2004 American Chemical Society

Distributed by Oxford University Press

PRINTED IN THE UNITED STATES OF AMERICA

Foreword

The ACS Symposium Series was first published in 1974 to provide a mechanism for publishing symposia quickly in book form. The purpose of the series is to publish timely, comprehensive books developed from ACS sponsored symposia based on current scientific research. Occasionally, books are developed from symposia sponsored by other organizations when the topic is of keen interest to the chemistry audience.

Before agreeing to publish a book, the proposed table of contents is reviewed for appropriate and comprehensive coverage and for interest to the audience. Some papers may be excluded to better focus the book; others may be added to provide comprehensiveness. When appropriate, overview or introductory chapters are added. Drafts of chapters are peer-reviewed prior to final acceptance or rejection, and manuscripts are prepared in camera-ready format.

As a rule, only original research papers and original review papers are included in the volumes. Verbatim reproductions of previously published papers are not accepted.

ACS Books Department

Contents

Application of Green Chemistry Principles in Agriculture

Indexes

Preface

As I compose this preface, I look out onto vast farmlands in central Illinois, fully aware that this scene is replicated worldwide. My eyes, moreover, have witnessed the amazing growth of green chemistry during the past eight years. Together, these experiences bring me to a unique perspective. We, as a civilization, are dependent upon agriculture for our very life, and the manner in which we practice agriculture must also be transformed under the new environmental paradigms emerging in green chemistry. Reciprocally, green chemistry can be inspired by much that nature (and agriculture) does in our world.

A match made in heaven, you say? Maybe not, but it is very important that there be a flow of information between the separate disciplines of green chemistry and agriculture. As a clarion of this fact, this volume can contribute to the process. The fact that a symbiotic relationship does already exist between these disciplines can be surmised from papers detailing research that documents it. The necessity of feeding our population, maintaining our environment, and practicing chemistry according to the new environmental mandate (green chemistry) explain why research in this field is escalating.

The symposium from which this book emerged began from discussions on what are the unique contributions that agriculture can make to the growing importance of green chemistry. It was not difficult to locate examples of present work in this interface between disciplines. For this glance into the exciting world of agricultural applications of green chemistry, researchers and workers from industry, academia, and government were selected to present papers during the ACS meeting, and to contribute their work to the present volume. What was apparent then, and is even more so now, is that this is merely the tip of the iceberg.

The book presents many facets of this interfacial, but yet seemingly integrated, enterprise. From fundamental studies on chlorophyll to pest and pesticide management, glimpses are taken of agriculture. Learning from agriculture (adhesives or remediation) form another pole in this work. The book is unusual and unique most probably because it **for the first time** proposes and demonstrates that green chemistry works to establish a path to sustainable agriculture.

The book will interest workers and researchers from green chemistry (for inspiration); scientists and educators in chemical and agricultural sciences (for an area of research that will be a leading wave); and industry and governmental leaders who will grasp the importance of this subject for the future.

But finally, I hope that the reader of this book will take what is read, add new ideas and insights, and perhaps contribute to this new area. Ultimately, this is how the ultimate goal of sustainability will be realized.

I acknowledge my own family (Millie, Maria, Milee, Liam, and Madeleine) for their daily support, for many of the contributors to this volume for being my mentors, and for Paul Anastas for being an inspiration. My work on this book and in this area has resulted from interactions with Tim Lindsey, Kishore Rajagopalan, the Pollution Prevention (P^2) group, and The Waste Management and Research Center (WMRC).

William M. Nelson
Waste Management and Research Center
1 Hazelwood Drive
Champaign, IL 61820–7465
(217) 244–5521 (direct)
(217) 333–8944 (fax)
wmnelson@uiuc.edu (email)

Green Chemistry and Agriculture

Chapter 1

Agricultural Applications in Green Chemistry

William M. Nelson

Waste Management and Research Center, 1 Hazelwood Drive, Champaign, IL 61820–7465

Agriculture is one of the oldest and global sources of human livelihood. It has matured from simple cultivation to sophisticated practices. Collectively, this complex situation exemplifies the sustainable agriculture dilemma.

From the symposium on "Agricultural Applications in Green Chemistry" (ACS, Orlando, 2001) and through this book we try to show that green chemistry offers an array of innovative approaches to agricultural practices and it looks for ways to accomplish more benign chemistry, through guidance by nature in agriculture. There is much to indicate opportunities for increased agricultural yield, economic benefits for manufacturers and end users, and enhanced environmental performance through this dynamic synergism.

Desirable qualities for agriculture

In a review chapter, "Green Chemistry and the Path to Sustainable Agriculture," Nelson delineates major desirable qualities for sustainable agriculture (reproduced in Table 1 below). Just as these are road signs on the path to sustainable agriculture, we can see how chapters in this book can fit nicely with these qualities. The characteristics can be used as a checklist of concerns regarding protection of the environment, production of healthy food and the practice of good ethics. The quality components have been placed into six categories. The protection of agricultural soils is essential for maintaining the production potential and ensuring a high quality of agricultural products. As agricultural activities affect not only the soil and agroecosystem, the protection of other biospheres, the atmosphere and groundwater must also be taken into consideration. Conservative resource practices are required to maintain our natural resources. The quality of agricultural products is affected by a wide range of production factors and by post-harvest procedures. Agricultural management also affects whether the appearance of landscape and countryside is attractive. Last but not least, our ethical view of nature determines how we evaluate and treat the agricultural milieu.

Table 1. Areas of concern for sustainableagriculture

Protection of agricultural soils

Soil erosion and salinization
Soil fertility
Subsoil compaction
Soil pollution

Protection of other biospheres, the atmosphere and groundwater

Use of pesticides
Leaching of plant nutrients
Emission of trace gases

Conservative resource practices

Use of water resources
Circulation of plant nutrients
Energy use
Biological diversity

High quality of agricultural products

Nutritiousness
Contamination
Hygiene

Attractive landscape and countryside

Appearance of the landscape
Appearance of the farm

Ethics

People
Livestock
Environment

Fundamental to any discussions of agriculture must be a current discussion of chlorophyll. Hoober and coworkers accomplish such a service in the chapter, "Chlorophylls b and c: Why do plants make them?" This serves not only a present need, but it also alludes to future areas of valuable research in the area of sustainable agriculture.

Going further in the understanding of this critical area Tripathy and coworkers discuss "Subplastidic distribution of Chlorophyll Biosynthetic Intermediates and Characterization of Protochlorophyllide Oxidoreductase C".

Natural product chemistry in agriculture

Pest management techniques have evolved over the past 50 years. Inorganic chemical pesticides were replaced by synthetic organic chemicals, and now biopesticides constitute a significant part of pest management technology. Kraus and coworkers give a lucid example of this new approach in the "Management of Soybean Cyst Nematode using a Biorational Strategy."

While conventional chemicals will remain as important pest management components, and the processes of combinatorial chemistry and high-throughput bioassays will allow the rapid synthesis and testing of large numbers of candidate compounds. New and equally important tools in pest management, with microbial pesticides and transgenic crops being likely to play important crop protection roles. Isman shows in the chapter "Plant essential oils as green pesticides for pest and disease management." there will be a continuing need for research-based approaches to pest control. His foundational work will be clear example to those who follow.

Weeds are known to cause enormous losses due to their interference in agroecosystems. Because of environmental and human health concerns, worldwide efforts are being made to reduce the heavy reliance on synthetic herbicides that are used to control weeds. Current reliance on pesticides also demands that we seek methodologies to properly remediate the lands. Larson and coworkers describe some of their work in this area in "Green Remediation of Herbicides: Studies with Atrazine."

As alternatives to existing control agents, a greener methodology is exemplified by the work of Wright and coworkers. In their paper, "Potential of Entomopathogenic Fungi as Biological Control Agents Against the Formosan Subterranean Termite," they give us a provative example.

Environmental Concerns

Misuse and incomplete understanding of the environmental fate of many industrial practices involving chemicals has resulted in environmental problems. Agriculture has been identified as the largest nonpoint source of water pollution, but it can also provide methodologies to even prevent pollution. In their contribution, "Agricultural green chemistry: in-process bioremediation of organic waste-containing aqueous solvents," Nelson and coworkers discuss the potential of bioremediation of organic wastes.

Green chemistry can also result by watching nature in the way it routinely works. Combie and coworkers have done this extremely well and they clearly describe the results in their contribution, "Adhesive Produced by Microorganisms."

Outlook

In a seminal article that stands out as a clear beacon, Rebeiz and coworkers clearly and confidently point toward the agriculture of the future. Their article, "Chloroplast Bioengineering: Photosynthetic Efficiency, Modulation of the Photosynthetic Unit Size, and the Agriculture of the Future," will be regarded as a truly insightful synthesis of experiment and theory.

This is the initial foray into the emergence of green chemistry leading to sustainable agriculture. With the wisdom and insight revealed by the scientists contained in this volume, and assured that the inspiration will continue along the path by attracting more scientists, I believe this is only the beginning of a wonderful and invaluable scientific enterprise.

Chapter 2

Green Chemistry and the Path to Sustainable Agriculture

William M. Nelson

Waste Management and Research Center, 1 Hazelwood Drive, Champaign, IL 61820–7465

The concept of sustainable agriculture is not new, and its desired characteristics are clearly stated. Green chemistry, though more novel, provides substance in scientific research and ultimately affords a path toward sustainable agriculture. The desired qualities and areas of focus in sustainable agriculture are delineated. These lead to perceived obstacles. These challenges offer to green chemistry novel possibilities for fruitful research utilizing its priciples and practices.

Introduction

Agriculture is one of the oldest sources of human livelihood and is found globally. It has developed from simple cultivation to sophisticated practices. In particular during the last century, mechanization, introduction of synthetic fertilizers and pesticides, and plant breeding have increased productivity and made crop production possible on previously uncultivated land. As a result, more humans can be fed. These changes created new kinds of problems for the environment and for society.

Environmental problems in agriculture vary from one country to another. Some of them are caused by natural conditions (high native heavy metal content, drought, volcanic eruptions, etc.), others depend on agricultural practices (leaching of nutrients and pesticides etc), and some are related to human influence in other areas (air pollution). Furthermore, these causes are often interrelated.

Collectively, this complex situation exemplifies the sustainable agriculture dilemma. Modern synthetic organic pesticides, fertilizers, herbicides, fungicides, and biocides are responsible for increasing the yield of agricultural production, decreasing human suffering, and enabling a world population of more than 6 billion people.[1] However, widespread application of agricultural chemicals has proven costly to the environment.

Green chemistry offers an array of innovative approaches to pest management, food production, and ecosystem protection. In this way it offers a more benign path to sustainable agriculture. There is much to indicate opportunities for increased agricultural yield, economic benefits for manufacturers and end users, and enhanced environmental performance. This book and the symposium from which it resulted provide a glimpse into the value and potential of green chemistry in sustainable agriculture.

Desirable qualities for agriculture

There are major desirable qualities for agriculture (Table 1).[2] The characteristics can be used as a checklist of concerns regarding protection of the environment, production of healthy food and the practice of good ethics. The quality components have been classified into six groups. The protection of agricultural soils is essential for maintaining the production potential and ensuring a high quality of agricultural products. As agricultural activities affect not only the soil and agroecosystem, the protection of other biospheres, the atmosphere and groundwater must also be taken into consideration. Conservative resource practices are required to maintain our natural resources. The quality of agricultural products is affected by a wide range of production factors and by post-harvest procedures. The whole life-cycle must be taken into consideration. Agricultural management also affects whether the appearance of landscape and countryside is attractive. Last but not least, our ethical view of nature determines how we evaluate and treat conditions.

Each of the concerns are directly or indirectly addressed by the 12 Principles of green chemistry.[3] Since a basic requirement for human survival is the sustainability of agriculture, how does Green Chemistry provide a path to realizing it? Discussions on the concept of sustainable agriculture have resulted in a certain consensus about four general aims: sufficient food and fibre production, environmental stewardship, economic viability and social justice. These are integrated into the concept of sustainable agriculture.[4] Green Chemistry charts a path to achieving sustainable agriculture by clarifying the paradigm and organizing principle: pollution prevention.

Table 1. Areas of concern for sustainable agriculture

Protection of agricultural soils

Soil erosion and salinization
Soil fertility
Subsoil compaction
Soil pollution

Protection of other biospheres, the atmosphere and groundwater

Use of pesticides
Leaching of plant nutrients
Emission of trace gases

Conservative resource practices

Use of water resources
Circulation of plant nutrients
Energy use
Biological diversity

High quality of agricultural products

Nutritiousness
Contamination
Hygiene

Attractive landscape and countryside

Appearance of the landscape
Appearance of the farm

Ethics

People
Livestock
Environment

Green chemistry applications in sustainable agriculture

This section describes in broad strokes some areas that are important for the development of our future agriculture. One may keep in mind that this selection is highly affected by the environmental conditions we live in and our personal knowledge. Implicit in every area is that green chemistry can play a pivotal role in accomplishing its goals.

Precision agriculture

Precision agriculture is a discipline that aims to increase efficiency in the management of agriculture. It is the development of new technologies, modification of old ones and integration of monitoring and computing at farm level to achieve a particular goal.[5] For example, the spatial variability of plant nutrients in fields affects the efficiency of nutrients added and thereby yield. Thus, techniques for recording variations within fields and the software to support the farmer when making decisions need to be developed. Prediction of mineralizable N in soils through combination of extraction methods with model simulation is one desirable way.[6] This will enable plant nutrients to be applied according to the nutrient status of the soil and the growing crop. Such precise application will optimize the utilization of manure and fertilizers and will help to increase yields and improve crop quality. Also, a spatially selective application of pesticides will help to reduce the amount of chemicals used.[7] Furthermore, methods to assess the N status of growing crops, for example via chlorophyll concentration in the tissue, are needed to avoid overfertilization with nitrogen and the resulting impact on N leaching.

Active management of soil biological processes

Soil loss from erosion annually removes up to 20 tons of soil per acre from lands under furrow irrigation. Scientists at the Northwest Irrigation and Soils Research Laboratory of the U.S. Department of Agriculture are exploring a polyacrylamide technology for reducing soil erosion. Mixing polyacrylamide with soil reduced sediment loss by an average of 94% in tests. By creating a water-soluble polyacrylamide solution that can be applied through irrigation systems, doses as small as 10 mg/L can be applied, corresponding to only 1–2 lb/acre compared with 500 lb/acre for dry application.[8] Improving soil retention also ensures that fertilizers and herbicides will remain on fields.

Maximum circulation of plant nutrients

Agricultural waste biomass may soon turn into a valuable resource for the production of chemicals and fuels. Biobased renewables have many advantages, such as reduced CO_2 production, flexibility, and self-reliance. This was also recognized by the chemical industry. For example, the Royal Dutch/Shell group estimated that by the year 2050 renewable resources could supply 30% of the worldwide chemical and fuel needs, resulting in a biomass market of $150 billion.[9]

Animal wastes

A balanced distribution of animal manure on farm areas is the most important step to establish effective circulation of plant nutrients. Furthermore, the development of new methods to handle and store solid animal manures on farms that enable nutrient conservation are desirable.

Food and urban wastes

Development of new or supplemental industrial systems for utilization of plant nutrients in municipal wastes is needed in order to enable recycling without contamination by environmental pollutants. Waste products need to be transported over longer distances to avoid too high nutrient levels in arable soils in the circumference of cities and towns. Methods that enable long-distance circulation are desirable.

Enzyme technology

Enzymes are being used in numerous new applications in the food, feed, agriculture, paper, leather, and textiles industries, resulting in significant cost reductions.[10] At the same time, rapid technological developments are now stimulating the chemistry and pharma industries to embrace enzyme technology, a trend strengthened by concerns regarding health, energy, raw materials, and the environment.

Enzymes are also used in a wide range of agrobiotechnological processes, such as enzyme-assisted silage fermentation, bioprocessing of crops and crop residues, fibre processing and production of feed supplements to improve feed efficiency. Especially the latter application, which includes the use of phytases to improve the efficiency of nutrient utilization and to reduce waste, is a rapidly growing sector.[11] The feed enzyme market now amounts to $150 million. In fact, 65% of poultry and 10% of swine feed already contain enzymes such as carbohydrases or phytase.

Several developments have started to tie in the agricultural sector with the chemical and pharmaceutical industries. Plants are being modified by genetic engineering for the production of polymers and pharmaceuticals such as antibodies or for improved nutritional value, for example, by increasing lysine or carotenoid content.[12] Only recently, ProdiGene announced the scale-up of trypsin production in recombinant plants, while other enzymes such as lactase may soon be produced in plants as well.[13]

Economics play a critical role in enzyme development. As an example, the price of cellulase needed to convert cellulosic biomass to fermentable sugars is a major factor. Therefore, the US Department of Energy awarded $32 million to Genencor and Novozymes to reduce the price of cellulose by a factor of ten, which could make bioethanol production and many other sugar-based fermentations economically viable.[14] A first technical milestone — the production of improved cellulase enzymes at one-half the cost of currently available technology — was reached in September 2001. Other issues pertinent to cellulose utilization include biocatalyst tolerance to acetate in the cellulose hydrolysate.[15] The biomass hydrolysis of lignocellulosic material to sugars would add a very significant market segment to the enzyme business: the potential cellulase market for available corn stover (leaves, stalks, and cobs), in the US Midwest alone is estimated to be $400 million, which would create the second largest enzyme market segment.

Natural product chemistry in agriculture

Pest management techniques have evolved over the past 50 years. Inorganic chemical pesticides were replaced by synthetic organic chemicals, and now biopesticides constitute a significant part of pest management technology. Requirements for the regulatory approval of pesticides changed dramatically in 1996 with the passage of the Food Quality Protection Act (FQPA). The FQPA directs the U.S. Environmental Protection Agency (EPA) to make more rigorous and conservative evaluation of risks and hazards and mandates a special emphasis on the safety of infants and children. Conventional chemicals will remain as important pest management components, and the processes of combinatorial chemistry and high-throughput bioassays will allow the rapid synthesis and testing of large numbers of candidate compounds. Biopesticides will become more important tools in pest management, with microbial pesticides and transgenic crops being likely to play important crop protection roles. There will be a continuing need for research-based approaches to pest control.[16] Once elucidated, these natural products become the targets of chemists specializing in synthesis.[17]

Weeds are known to cause enormous losses due to their interference in agroecosystems. Because of environmental and human health concerns, worldwide efforts are being made to reduce the heavy reliance on synthetic herbicides that are used to control weeds. In this regard the phenomenon of allelopathy, which is expressed through the release of chemicals by a plant, has been suggested to be one of the possible alternatives for achieving sustainable weed management.[18] The use of allelopathy for controlling weeds could be either through directly utilizing natural allelopathic interactions, particularly of crop plants, or by using allelochemicals as natural herbicides. The allelochemicals present in the higher plants as well as in the microbes can be directly used for weed management on the pattern of herbicides. Their bioefficacy can be enhanced by structural changes or the synthesis of chemical analogues based on them.

Green chemistry and sustainable agriculture: future challenges

The thin layer of soil on the earth's surface performs many functions essential to life.[19] Sustainable agriculture focuses on soil issues. Soils research has accomplished much, providing us with a thorough understanding of the physical, chemical, and biological properties and processes of soils, determining the role of soils in environmental quality, and developing the management practices used to produce a bountiful food supply. However, despite these accomplishments and continued demands for soils-related information, soil scientists are currently facing many challenges. A steady supply of inexpensive, high quality food produced by less than 2% of a largely urban population has left the majority of people with little appreciation of the problems and challenges facing agriculture. Soil scientists must ensure that the science is available to address critical problems facing society, namely: population pressure and the need for increasing agricultural productivity; competing uses for land and water resources; dependence on nonrenewable resources; and environmental quality, especially in developing countries. Facing current challenges and solving future problems will likely require that soil scientists conduct research differently than in the past, with greater emphasis on holistic team- and interdisciplinary analyses of problem areas. This moves into the heart of green chemistry.

Future challenges

The challenges facing soil science and agriculture concern managing the soil resource to insure that the functions performed by soils are maintained and societal demands are met. Not only is the human population increasing rapidly, there is a desire in most societies, and especially in developing countries, for our standard of living to improve. This implies that not only will there be more people to provide for but that those people will be expecting a higher level of goods and services. As we strive to meet these demands we will be required to develop management practices and utilize resources in such a way that the resources will be available to perform the functions and meet the needs of future generations.

Population Pressures

Recent estimates suggest that there will be an additional 1 billion people on earth within a decade.[20] Although there is little doubt that the human population is increasing, the rate at which this change is occurring sheds light on the demands that will be placed on production agriculture in the near future. Changes in reproductive rates have decreased throughout much of the world. In developed countries, many couples are having two or fewer children, whereas in developing countries, the reproductive rate is often much greater. In addition, longevity has increased the life expectancy in developed countries to a greater extent than in developing countries. The combination of lower reproductive rate and longer life expectancy has resulted in an aging population that is increasing slowly in many developed countries. In contrast, the age structure in many developing countries resembles a pyramid, with a large percentage of the population in younger age classes. By the year 2020, it is expected that the world population will exceed 8 billion people, and more than 80% of these people will live in developing countries.

Need for Increasing Production

In our current state of grain surpluses and low commodity prices, it is difficult to appreciate that the current rate of increases in grain production are below those needed to supply food for the human population in the relatively near future (next 30 years). As a result of the population trend given above, grain yields will have to increase from 1.2% (for wheat and rice) to 1.5% (for corn) per year to meet future demand.[21]

Our knowledge of the effect of water (e.g., yield as a function of water availability) and nutrient availability on crop performance is relatively good. Further improvements in management will likely come about through improved understanding of more subtle and complex concepts such as:

- nutrient-disease interactions,
- soil-water interactions
- crop rotation effects,
- changes occurring when management practices change, and
- others unimagined.

Two disciplines that offer great potential for contributing to improved management are plant breeders and soil microbiologists. Strong interaction between soil scientists and plant breeders in identification of stresses, selection of varieties tolerant of specific stresses, and development of management practices to minimize the effect of stresses to which varieties are most susceptible would improve crop performance and breeding efforts greatly. Soil biology is the least understood and most underutilized area of soil science. The potential for improving nutrient availability and utilization, managing pest organisms, and ameliorating degraded soils is largely unknown.

Environmental Concerns

Inorganic fertilizers, synthetic pesticides, and other agrochemicals have played an essential role in increasing efficiency and productivity in modern agriculture. Misuse and incomplete understanding of the environmental fate of these chemicals has resulted in environmental problems. Agriculture has been identified as the largest nonpoint source of water pollution. Nutrient enrichment of estuaries along the Atlantic coast has been suggested as the cause of outbreaks of pfisteria. States have reported that 40% of the waters they have surveyed are impaired for recreational and wildlife uses.[22] As discussed earlier, our understanding of processes and reactions in soils has improved and management practices that increase efficient use of chemicals and minimize negative environmental impacts have been developed. Public pressure will doubtlessly require further progress in this area. As agricultural pressures increase, further efforts in this area will be needed. Soil scientists, pesticide chemists, representatives from the fertilizer industry, hydrologists, microbiologists, and others will have to interact to maintain and improve agricultural productivity and environmental quality.

The potential of soil biology for improving nutrient use efficiency, control of soil borne pests, remediation of contaminated soils and water, and reducing greenhouse gas emissions is largely unknown. Use of biotechnology in soil biological research suggests that the vast majority of soil microorganisms have yet to be identified. The potential that exists for bioengineering to manage soil biota and mediate soil processes is also largely unknown, mainly because of the challenge of matching organisms to a field or soil environment in which they are active and their traits expressed.

Dependence on Fossil Fuels

Interception of solar energy constitutes, by far, the largest energy input to agriculture. Mechanization, increased use of inorganic fertilizers and synthetic pesticides, increased use of irrigation, and on-farm practices such as grain drying have increased the total amount of fossil fuel used in agriculture. Six percent of total energy use in the United States is in the field of crop production.[23] Livestock manure is currently treated more as a waste than a fertilizer and C source. Management practices that better utilize nutrients in manure, crop residues, and cover crops need to be developed. In addition, solar, wind, and biofuels technologies will have to be developed to reduce our dependence on fossil fuels.

Soil Degradation

Slightly more than 3 billion of the earth's 13 billion ha of land area has the potential for use as agricultural land.[24] Approximately 50% of the potentially arable land is currently in arable or permanent crops. An additional 2 billion ha has been degraded or destroyed, largely through mismanagement.[24] Land continues to be degraded by erosion, salinization, and waterloggging at a rate of 10 million ha per year.[25] Nearly 80% of the potentially cultivable land in developed countries is currently under cultivation compared with less than 40% in developing countries. This suggests that as the demand for agricultural goods increases, use of new land for crop production will expand most rapidly in the developing countries. Many areas where expansion will occur possess soils of lower productivity and higher susceptibility to degradation. A major challenge will be the development of agricultural practices that optimize productivity and maintain soil quality in marginally productive soils in these developing countries.

Scientists who understand the functions of soil as a natural body will need to interact with scientists of other disciplines and members of development agencies as agriculture expands to ensure that soil and water resources are maintained and used in a sustainable manner. Existing knowledge of processes that degrade soils and the effect of management practices on soil functions will need to be incorporated with regional cultural and economic conditions to develop sustainable management practices.

Outlook

The production of food has to increase as the global human population will increase by about two billion during the next 25 years.[26] Thus, intensive production seems absolutely necessary to guarantee that production will be able to keep pace with population growth. The critical question is whether it will be possible to increase production without an increase or even a lowering of emissions. In general, emissions from arable land increase with more intensive fertilization. Nitrate leaching, for example, increased slightly with higher fertilization intensity and first at an excessive supply of N fertilizer, leaching reached very high and unsatisfactory levels.[27] Addiscott[28] pointed out that low-intensive crop production is least sustainable, whereas high-intensive use of arable land is most sustainable, in accordance with the theory of thermodynamics. Furthermore, with high yields per area, more food can be produced and more land can be saved for other uses. This is most important in countries with limited land resources and a high population density. Still, a high degree of knowledge is essential for intensive agriculture to be able to utilize the means of production in a highly efficient way and avoid misuse of resources, overfertilization and any negative effects on the environment.

Nutrient imbalances

Regional specialisation of farms has resulted in production that is most often much greater than the need of the immediate market. Agricultural products are transported long distances, both crops used for human consumption and fodder concentrates for animal husbandry, which means a net removal and no return of harvested nutrients. On the other hand, a large import of feeding stuff to farms contributes to an excessive supply at a local or even regional level. This more or less open plant nutrient cycle causes nutrient imbalances. For example, concentrates may be produced on land in developing countries where rain forests were cut and soils may degrade through erosion and nutrient depletion.

One probable way to affect farming in the future is through analysis and classification of agricultural production and environmental stewardship on individual farms. This analysis may stimulate favourable farming development. If properly designed and well founded, a quality assessment system for agriculture can be a driving instrument to be used on the path to sustainable agriculture.

Are present nutrient management recommendations for the world's major cereal cropping systems adequate to sustain the productivity gains required to meet food demand while also assuring acceptable standards of environmental quality? Because average farm yield levels of 70-80% of the attainable yield potential are necessary to meet expected food demand in the next 30 years, research must seek to develop nutrient management approaches that optimize profit, preserve soil quality, and protect natural resources in systems that consistently produce at these high yield levels. Significant advances in soil chemistry, crop physiology, plant nutrition, molecular biology, and information technology must be combined in this effort.[29]

With increasing concerns about the environment, better use of the natural resource base, less use of chemicals and efficient use of irrigation water have become increasingly important goals of sustainable agriculture.[30] Use of biofertilizers offers agronomic and environmental benefits for intensive agricultural systems.

A quality assessment system for agriculture

On the path to sustainable agriculture through green chemistry the influence of agricultural practices on the environment, the status of selected properties and the efficiency of production must be taken into account. The areas of concern outlined earlier can be useful for a structural outline. A comprehensive quality assessment system, combining different aspects of production and environmental stewardship, can be a very powerful tool to direct development towards environmentally sound and sustainable agriculture.

The use of such a system may favor agricultural production in certain areas and question it in others. Within a country, this assessment may lead to setting aside agricultural land. However, as food production is a fundamental need for humans, most nations are interested in producing their own food to some extent. The result could be that agricultural land used in one country could be set aside in another. Where to actually carry out agriculture is therefore also a political decision.

Conclusions

On a global scale, we need to increase food production and at the same time ensure the quality of agricultural soils and of the surrounding environment. This article has attempted to link green chemistry and sustainable agriculture to reach the goals of sufficient food production and environmental stewardship. An awareness and application of these quality components is useful to gain an overview of the conditions of agriculture and they are also considered as guidelines for agricultural research and development. Innovation, creative solutions and discoveries based on natural sciences will be helpful in the development of sustainable agriculture.

To address this problem, new approaches are needed, and particularly for pest control and the agricultural chemicals industry, green chemistry may provide opportunities.[3,31,32,33,34,35,36] Green chemistry approaches follow a growing trend in industry, motivated by simultaneous requirements for environmental improvement, economic performance, and social responsibility.

Clearly, having to make a choice between sufficient food and clean water and ecosystem survival is not acceptable.[37] The issue, however, is not the need to make a choice between protecting crops critical to human sustainability and a healthy environment. Rather, it is a challenge to the world chemical, biological, and agricultural communities to devise new methods to protect and enhance plant growth and yield while eliminating downstream consequences. This is why green chemistry is important. It provides tools to protect environmental quality in the face of increasing global pressures on food production. (U.S. EPA's annual Presidential Green Chemistry Challenge Award:[37] Biomimetic approaches,[38] Toward less fertilizer, New impacts from biotechnology.[39])

References

(1) Waichman, A. V.; Rombke, J.; Ribeiro, M. O. A.; Nina, N. C. S. *Environ. Sci. Poll. Res.* **2002**, *9*, 423-428.

(2) Kirchmann, H.; Thorvaldsson, G. *Europ. J. Agronomy* **2000**, *12*, 145-161.

(3) Anastas, P. T.; Warner, J. C. *Green Chemistry: Theory and Practice*; Oxford University Press: New York, 1998.

(4) Lowrance, R.; Hendrix, P. F.; Odum, E. P. *Am. J. Altern. Agric.* **1986**, 1169-1173.

(5) Blackmore, S. *Outlook Agric* **1994**, *23*, 275–280.

(6) Appel, T.; Mengel, K. *Zeitschrift fur Pflanzenerna"hrung und Bodenkunde* **1998**, *161*, 433–452.

(7) Stafford, J. V.; Miller, P. C. H. *Comput. Electron. Agric.* **1993**, *9*, 217–229.

(8) EPA, U. S. "The Presidential Green Chemistry Challenge Awards Program: Summary of 1998 Award Entries and Recipients," U.S. Government Printing Office, 1998.

(9) OECD. "Biotechnology for Clean Industrial Products and Processes," OECD, 1998.

(10) van Beilen, J. B.; Li, Z. *Current Opinion in Biotechnology* **2002**, *13*, 338-344.

(11) "Annual Report," Novozymes AS, 2001.

(12) Fraser, P. D.; Roemer, S.; Kiano, J. W.; Shipton, C. A.; Mills, P. B.; Drake, R.; Schuch, W.; Bramley, P. M. *J Sci Food Agric* **2001**, *81*, 822-827.

(13) Hood, E. E. *Enzyme Microb Technol* **2002**, *30*, 279-283.

(14) Russo, E. *The Scientist* **2001**, *15*, 1-4.

(15) Lawford, H. G.; Rousseau, J. D.; Tolan, J. S. *Appl Biochem Biotechnol* **2001**, *91*, 133-146.

(16) Wheeler, W. B. *Journal of Agricultural & Food Chemistry* **2002**, *50*, 4151-4155.

(17) Crombie, L. *Pestic Sci* **1999**, *55*, 761 - 774.

(18) Singh, H. P.; Batish, D. R.; Kohli, R. K. *Critical Reviews in Plant Sciences* **2003**, *22*, 239-311.

(19) Wienhold, B. J.; Power, J. F.; Doran, J. W. *Soil Science* **2000**, *165*, 13-30.

(20) Haub, C.; Farnsworth-Riche, M. In *Beyond the Numbers*; Mazur, L. A., Ed.; Island Press: Washington, DC, 1994, pp 95-108.

(21) Rosegrant, M. W.; Leach, N.; Gerpacio, R. V. "Alternative Futures for World Cereal and Meat Consumption," International Food Policy Institute, 1998.

(22) Survey, U. S. G. "The quality of our nation's water-Nutrients and pesticides," Dept. of the Interior, 1999.

(23) Evans, L. T. *Feeding the Ten Billion: Plants and Population Growth*; Cambridge Univ. Press: Cambridge, UK, 1998.

(24) Lal, R. *Adv. Soil Sci.* **1990**, *11*, 129-172.

(25) Pimentel, D., C. ; Harvey, P.; Resosudarmo, K.; Sinclair, D.; Kurz, M.; McNair, S.; Crist, L.; Shpretz, L.; Fitton, R.; Saffouri; Blair, R. *Science* **1995**, *267*, 1117-1123.

(26) Greenland, D. J.; Gregory, P. J.; Nye, P. H. "Land Resources: On the Edge of the Malthusian Principle," CAB International, 1997.

(27) Bergstrom, L.; Brink, N. *Plant Soil* **1986**, *93*, 333–345.

(28) Addiscott, T. M. *Eur. J. Soil Sci.* **1995**, *46*, 161–168.

(29) Dobermann, A.; Cassman, K. G. *Plant and Soil* **2002**, *247*, 153-175.

(30) Monem, M. A. S. A.; Khalifa, H. E.; Beider, M.; El Ghandour, I. A.; Galal, Y. G. M. *Journal of Sustainable Agriculture* **2001**, *19*, 41-48.

(31) Anastas, P. T.; Farris, C. A., Eds. *Benign by Design: Alternative Synthetic Design for Pollution Prevention*; Oxford University Press: New York, 1994.

(32) Anastas, P. T.; Tundo, P., Eds. *Green Chemistry: Challenging Perspectives*; Oxford University Press: New York, 2000.

(33) Anastas, P. T.; Williamson, T. C., Eds. *Green Chemistry: Designing Chemistry for the Environment*; Oxford University Press: New York, 1996.

(34) Anastas, P. T.; Heine, L. G.; Williamson, T. C., Eds. *Green Chemical Syntheses and Processes*; American Chemical Society: Washington, DC, 2000.

(35) Anastas, P. T.; Williamson, T. C., Eds. *Green Chemistry: Frontiers in Benign Chemical Synthesis and Processes*; Oxford University Press: New York, 1998.

(36) Hjeresen, D. L.; Anastas, P. T.; Kirchhoff, M.; Ware, S. *Environ. Sci. Technol.* **2001**, *35*, 114 A-119A.

(37) Hjeresen, D. L.; Gonzales, R. *Environ. Sci. Technol.* **2002**, *36*, 102 A-107 A.

(38) Knipple, D. C.; Rosenfield, C.-L.; Miller, S. J.; Liu, W.; Tang, J.; Ma, P. W. K.; Roelofs, W. L. In *Proc. Natl. Acad. Sci. U.S.A.*, 1998; Vol. 95, pp 15,287-215,292.

(39) Wei, Z.; Laby, R. J.; Zumoff, C. H.; Bauer, D. W.; He, S. Y.; Collmer, A.; Beer, S. *Science* **1992**, *257*, 85-88.

Chapter 3

Chlorophylls *b* and *c*: Why Plants Make Them

Laura L. Eggink, Russell LoBrutto, and J. Kenneth Hoober[*]

School of Life Sciences and Center for the Study of Early Events in Photosynthesis, Arizona State University, Tempe, AZ 85287–4501

Stability of light-harvesting complexes in plants requires synthesis of chlorophyll (Chl) *b* by oxidation of the 7-methyl group of Chl *a* to the 7-formyl group. The electron-withdrawing property of the formyl group redistributes the molecular electron density toward the periphery of the macrocycle, away from the central Mg, which increases the Lewis acid 'hardness' of the metal. Consequently, stronger coordination bonds are formed with 'hard' Lewis-base ligands, such as carboxyl groups and the oxygen-induced dipole of amide groups, ligands that are relatively unfavorable for coordination with Chl *a*, in the apoproteins of the light-harvesting complexes. Chl *a* is oxidized to Chl *b* by Chl *a* oxygenase. The EPR spectrum of the enzyme revealed signals for a mononuclear iron and an iron-sulfur complex, which are predicted by the amino acid sequence, and also for a stable radical on an amino acid, possibly a tyrosine. Chl *c,* in which conjugation of the ring π system extends through the double-bond of the *trans*-acrylate side-chain to the unesterified, electronegative carboxyl group, serves the same role in chromophytic algae as Chl *b* does in plants.

The technological advances of Green Chemistry are critically important for preservation of our environment and conservation of resources. Equally important, and perhaps more so, is understanding the impact of chemicals on the environment. The most fundamental process that must be protected is photosynthesis in plants, which supports all living organisms. The key process in photosynthesis is activation of chlorophyll (Chl) as a reducing agent by the absorption of light energy. The efficiency of this process is dramatically enhanced by light-harvesting antennae, which increase the absorptive cross-section of photosynthetic units. Thus, the enzyme-catalyzed oxygenation of Chl *a* to Chl *b*, which is required for assembly of the light-harvesting complexes (LHCs), is among reactions of great importance.

Photosynthetic organisms, from cyanobacteria to plants, contain Chl *a*, which is the principal Chl in photosynthetic reaction centers. Plants and green algae also contain Chl *b*, with a Chl *a:b* ratio of 2.5 to 4. Chromophytic algae, such as brown algae, diatoms and dinoflaggellates, contain Chl *c* rather than Chl *b* (see Figure 1 for structures). Chl *b* and Chl *c* are found only in LHCs (*1,2*). Because of their location in LHCs, the function of Chls band *c* has traditionally been ascribed to enhancement of light absorption, because of their slightly shifted absorption spectra relative to Chl *a*. However, this function does not explain the dramatic reduction in the amount of LHCs in mutant plants that lack the ability to synthesize Chl *b*. Hoober and Eggink (*3*) and Eggink et al. (*4*) proposed that changes in the coordination properties of the central metal atom, caused by substituents on the tetrapyrrole ring (*5-11*), affect LHC assembly and thereby dramatically influence the size of the light-harvesting antennae.

Coordination Chemistry in LHCs

Interaction of Chl with proteins involves formation of coordination bonds between the Mg atom of Chl as the Lewis acid and amino acid side-chains as Lewis bases. To understand the role of Chl *b* in LHC assembly, the effects of chemical changes at the periphery of the tetrapyrrole on these interactions must be considered. Early steps in the synthesis of Chl from Mg-protoporphyrin IX methyl ester involves formation of the fifth, isocyclic ring, which generates the electron-withdrawing, 13^1-carbonyl group, and reduction of the C17-C18 double bond. Intermediates in the biosynthetic pathway include precursors that contain electron-withdrawing vinyl groups on positions 3 and 8 or the reduced, electron-donating ethyl group on position 8 (*7,8*). Chl *a* occurs most commonly as the mono-vinyl form, shown in Figure 1. Subsequent oxidation to Chl *b* replaces the electron-donating 7-methyl group on Chl *a* with the electronegative formyl group. Introduction of this electron-withdrawing group of Chl *b* causes a further redistribution of electrons away from the central pyrrole nitrogens, which reduces their electron density and lowers their pK values (*5,6*). The lower electron density results in less shielding of the Mg atom, which consequently expresses a stronger positive point charge (*9,10*). Chl *c* is synthesized from protochlorophyllide, the precursor of Chls *a* and *b*, not by reduction of the double bond between C17 and C18 but by introduction of a double bond in the side-chain to form the trans-acrylate group (Figure 1). The acrylate carboxyl group remains unesterified, in contrast to other Chls, and as a result, conjugation of the ring π system is extended to the electronegative carboxyl group, which also lowers the electron density around the Mg atom (*11*). Although by a different route, the Mg atom in Chl *c* thus achieves characteristics similar to those in Chl *b*.

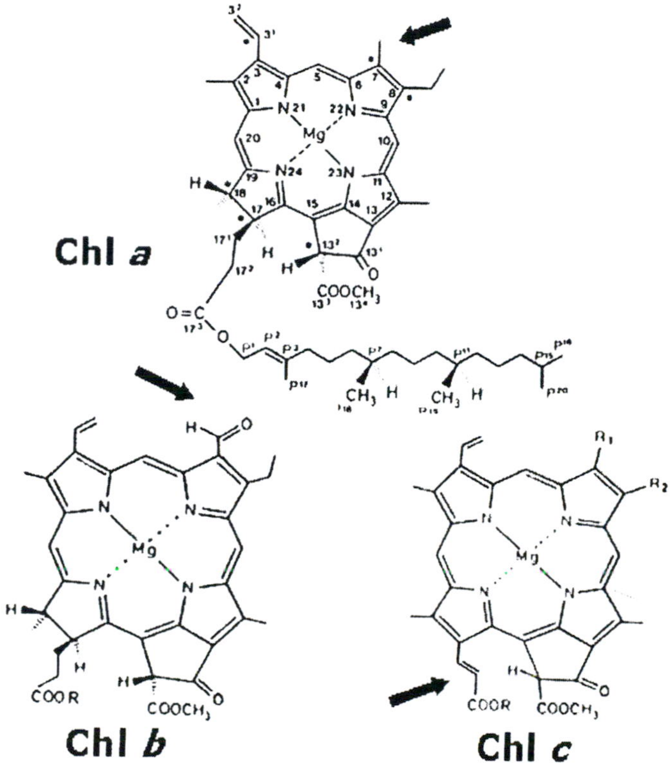

Figure 1. Structures of mono-vinyl Chls a, b and c₁. Distinctive differences in each structure are indicated by arrows. Chl a contains a 7-methyl group, which in Chl b is oxidized to a formyl group. Chls a and b are otherwise the same, including esterification of the C17 propionyl group with the isoprenoid alcohol (R in Chl b = phytol). The C17-C18 double-bond in ring D of Chl c is not reduced, and an additional double-bond is introduced into the side-chain to produce the trans-acrylate group, with R = H. R₁ and R₂ in Chl c₁ are methyl and ethyl groups, respectively.

The Mg in Chls *b* and *c,* with a more positive point charge, should consequently interact more electrostatically with 'hard' Lewis bases that contain an electronegative oxygen-induced dipole. For example, Chl *b* binds water much more strongly than does Chl *a* (*12*) and should also coordinate strongly with oxygen atoms in carboxyl groups and the carbonyl group of amides, including peptide bonds. On the other hand, the greater electron density around the Mg in Chl *a* should repel such ligands. More likely, Chl *a* forms coordination bonds by orbital interaction, which favors complexes with the imidazole ring of histidine. With few exceptions, occupancy of Chl binding sites in LHCII after reconstitution is consistent with this concept of ligand selectivity (*13-15*). Because of its stronger interaction with water, the very slow on-rate for Chl *b* binding to the protein during reconstitution (*16*) may allow Chl *a* to occupy sites that, based on data from *in viva* studies, should be filled with Chl *b* (*3,4*). Nevertheless, reconstitution of the complex *in vitro* occurs with a very high degree of selectivity in each Chl binding site (*13,15*).

The specificity of binding appears to reflect equilibria between interacting species. Tamiaki et al. (*17*) showed, with an insightful series of experiments, that introduction of an oxygen atom on the periphery of Zn-tetrapyrrole molecules increased the equilibrium constant for a complex with pyridine as the ligand in benzene. A Chl *b* analogue provided an equilibrium constant for a 1:1 complex that was nearly two-fold greater than the complex with a Chl *a* analogue. These data were interpreted as an increase in the Lewis acid strength of the metal caused by introduction of oxygen atoms onto the ring system.

Chl *b* provides greater stability to LHCs through coordination with ligands that are unfavorable for coordination with Chl *a* (*4*). This concept implies that Chl plays an *active* role in assembly of LHCs, with Chls *a* and *b* selecting different ligands. This requirement of Chl for folding of the apoprotein (LHCP) was demonstrated during reconstitution studies (*18*). Rather than the apoprotein simply providing generalized ligands for the pentacoordinate Mg in Chl, modification of the Chl molecule generates selectivity in ligand binding and expands the ligand selection beyond the imidazole ring of histidine, the favored ligand for Chl *a,* to Lewis bases containing an electronic dipole. Consequently, stable LHCs accumulate and the capacity of the plant to harvest light energy is greatly increased by expansion of the antenna. LHCPs are required for accumulation of Chl *b,* although it is not clear whether oxidation of Chl *a* occurs after the Chl molecule has bound to the protein but before folding of the complex is completed *in viva,* or whether LHCPs are regulatory effectors of the synthesis of Chl *b.*

Chlorophyll Assignments in LHCII

LHCII contains 12 to 14 Chl molecules, 7 or 8 of Chl *a* and 5 or 6 of Chl *b.* Figure 2 shows the proposed binding sites and ligands for 12 of the Chls. According to the concept of ligand selectivity noted above, the histidine imidazole groups in sites *a*5 and *b*3 should be ligands for Chl *a*. The peptide backbone carbonyl in site *a*6, amides in sites *a*3 and *b*6 and a carboxyl group (possibly in an ion-pair with an adjacent guanidinium group) in *b*5 were identified as ligands to Chl *b* by reconstitution studies (*13-15*). Chl *b* has been proposed to fill site *b*2, although a ligand has not been identified in this site. Because binding of Chl *b* to helix-1 of the apoprotein during import into the chloroplast is required to retain the protein in the organelle (*3,4*), we propose that the glutamate carboxyl group that eventually provides site *a*4 must initially bind Chl *b*. The atomic structure of the final complex indicates that this carboxyl group forms an ion-pair with the strongly positive-charged guanidinium group of an arginine residue on helix-3 (*19*). Following reconstitution, these residues form a ligand for Chl *a*. it is not inconceivable that folding of the protein, with formation of the ion-pair as helix-3 approaches helix-1, creates a ligand more favorable for Chl *a*, which consequently results in displacement of the Chl *b* molecule. A similar scenario could occur with the glutamate in site *a*1, which initially may serve as a ligand to Chl *b* but switches to Chl *a* as rapprochement of the guanidinium group of helix-1 forms the ion-pair. Thus, Chl *b* in site *b*2, and perhaps *b*1, may result from a change in Lewis base character and thus shift the equilibria of interactions in sites *a*1 and *a*4 to favor Chl *a*, as folding of the complex is completed. This possibility is supported by the switch from a mixed site containing Chl *a* in *b*6 in CP29 (apoprotein Lhcb4), in which a glutamate is paired with a ion, to Chl *b* when the glutamate is replaced with a glutamine residue (*20*). CP29, a minor LHCII, contains only two Chl *b* molecules, most likely in sites *b*5 and *a*3. A highly excitonically-coupled heterodimer was detected in CP29 that included a blue-shifted Chl *b* that absorbs near 640 nm (*21*), which is probably Chl *b* in *a*3 paired with Chl *a* in *b*3. In reconstitution studies, the central sites designated *a*1, *a*2, *a*4 and *a*5 were filled specifically with Chl *a* (*13*).

Data are available with *in vivo* systems that indicate that the primary role of Chl *b* is to facilitate assembly of stable Chl-protein complexes (*22-24*). LHCPs are not imported into the chloroplast at a significant rate in the absence of Chl synthesis (Figure 3). The proteins are synthesized but accumulate instead in the cytosol and/or vacuoles (*25*). The fact that they accumulate as the mature-sized proteins nevertheless suggests that entry into the chloroplast was sufficient for processing to remove the N-terminal targeting sequence. Even under rapid synthesis of Chl and assembly of LHCs, LHCPs are made in excess (*26*). The unused LHCPs are partially imported into the chloroplast and then retracted and degraded in the cytosol and/or vacuoles. Mutant strains of plants and algae that lack Chl *b* are markedly deficient in LHCs but not in synthesis of the apoproteins (*25,27,28*). The major LHCPs in these mutants are mature-sized and thus again appear to enter the chloroplast sufficiently for N-terminal processing and then are retracted into the cytosol and degraded. A high rate of Chl *a* synthesis can partially compensate for the absence of Chl *b* and allow LHCP accumulation in thylakoid membranes (*25,28,29*). Chl *a* can probably fill most binding sites in the protein, although stronger coordination bonds are formed in

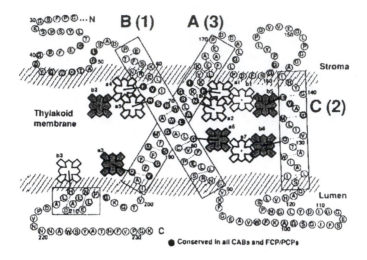

Figure 2. Proposed binding sites for 12 Chl molecules in LHCII. Chl b is designated as the filled tetrapyrrole symbols. The assignments were based on data in references 13-15. The most probable mechanism of LHCII assembly suggests that binding of Chl b to site a4 in helix-I is essential for accumulation of the complex (3,4). As indicated in the text, completion of folding may alter ligand character in sites a1 and a4 to allow displacement of Chl b by Chl a, the final occupant of these sites. Additional Chls, not bound to the protein through amino acid side-chain ligands, may occur in the complex in viva. The model for the structure of LHCII was adapted from reference 1.

some sites with Chl *b*. in contrast to LHCPs in complexes containing Chl *a* and *b*, those in complexes containing only Chl *a* are readily degraded when thylakoid membranes are incubated with proteases (*22,24*).

Cellular Location of Chl *b* Synthesis and LHCII Assembly

Observations regarding LHCII assembly can be correlated with the sub-chloroplast location of Chl *b* synthesis. Immunoelectron microscopy was used to determine the distribution of LHCPs in cells of *Chlamydomonas reinhardtii in vivo*. In cells grown in the light, antibodies against LHCPs extensively decorated thylakoid membranes (*4,30*). However, most strikingly, in the mutant strain *cbn*l-113*y*, which is unable to synthesize Chl *b* and also does not make Chl *a* in the dark, LHCPs were synthesized at nearly the same rate in the dark as in the light (Figure 3, *left panel*) (*31*) but were not detected immunochemically in the chloroplast (Figure 3, *right panel*). Instead, the proteins accumulated in the cytosol and vacuoles. Chl *b*-less cells grown in the light with CO_2 as the carbon

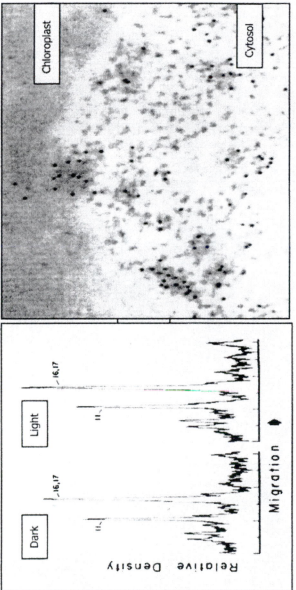

Figure 3. (left panel). Cells of C. reinhardtii cbn 1-113y, a Chl b-minus strain, were labeled with [¹⁴C]arginine in the dark or light at 38°C for 15 min. A pellet, membrane-containing fraction of broken cells was prepared and the proteins were resolved by electrophoresis. Radioautograms of gels were scanned to estimate the rate of synthesis of the major LHCPs, designated polypeptides 11, 16 and 17. (right panel) A section of a cell incubated in the dark at 38°C was stained with antibodies to polypeptide II and then protein A-gold to localize bound antibodies (25).

source were chlorotic and contained fewer LHCPs in the chloroplast than in cells of the same strain grown with acetate as a carbon source (25).

During chloroplast development, induced by exposure of degreened cells to light at 38°C, LHCPs were initially detected in the chloroplast along the envelope (Figure 4). The linear accumulation of LHCPs correlated with the immediate and nearly linear accumulation of Chl *b* (32). The proteins accumulated also at a similar rate in polyphosphate granules in small vacuoles in the cytosol (26). Thus, LHCPs made in excess of the chloroplast's capacity to assemble LHCs were not transported into the chloroplast stroma but instead were shunted to vacuoles for degradation. In a mutant strain of *C. reinhardtii* designated MC9, which is deficient in assembly of LHCs, LHCPs were shunted to vacuoles during chloroplast development at a rate nearly 10-fold greater than in wild-type (33). Regardless of whether most of the proteins accumulated within or outside of the chloroplasts, electrophoretic analysis indicated that all detectable proteins were mature-sized. These results show that LHCPs interact with Chl in the chloroplast envelope, after partial import to allow processing to

Figure 4. Immunoelectron microscopic localization of LHCPs after 15 min of greening of yellow cells of C. reinhardtii yl. A section of the cell was treated with antibodies against LHCP and subsequently with protein A-gold to localize the bound antibodies. Gold particles were detected predominantly over the areas inside, but immediately adjacent to, the chloroplast envelope. C, chhoroplast; ce, chloroplast envelope. (From ref. 26).

remove the N-terminal targeting sequence, and that Chl *b* (or Chl *c*) is required to prevent the proteins from escaping from the chloroplast envelope back into the cytosol. These results provide indirect evidence that Chl *b* is synthesized in the envelope, where it immediately interacts with LHCPs. The final steps in synthesis of the unesterified chlorophyllide *a* occur in the envelope (*34,35*) and the chlorin is apparently immediately esterified with geranylgeranyl diphosphate (*32*). More direct evidence for localization of these activities are given below.

Kinetic analyses of the assembly of LHCs, and their connection to an energy transducing apparatus that traps light energy absorbed by LHCs, showed that functional membranes were assembled within minutes after degreened cells were exposed to light (*26*). Activities of photosystems I and II appeared with no lag under these conditions (*36*). These results indicate that complete and functional photosynthetic membranes are assembled rapidly, within the time expected for synthesis of a LHCP in the cytosol, import into the chloroplast and assembly into a complex with Chl. Although these observations have been made most definitively with studies with the model alga *C. reinhardtii*, consistent data have been obtained with plants (see refs. *24,37* for reviews). Mutant strains of the plant *Arabidopsis* (*38*) and the cyanobacterium *Synechocystis* (*39*) lack an activity necessary for generation of vesicles from the inner envelope membrane, which is necessary for transfer of material to the interior of the chloroplast for expansion of the thylakoid membrane system.

The 'Retention Motif'

A generalized sequence motif—ExxHxR—in the membrane-spanning helix-1, the first of the Chl binding domains to enter the envelope membranes during import into the chloroplast, has been conserved throughout evolution from small, single-membrane-spanning polypeptides in cyanobacteria to all LHCPs and related proteins in plants (*1,40,41*). In a few of these proteins, H (histidine) is replaced with another amino acid, N (asparagine). The latter motif is also located in helix-3 of LHCPs as —ExxNxR— (*1,24*). Molecular modeling of the amino acid sequence in helix-1 indicated that the peptide should provide two ligands for Chl *a*, one provided by an ion-pair formed between the carboxyl group of F (glutamate) and the guanidinium group of R (arginine) and the second by the imidazole ring of histidine. Furthermore, in the amino acid sequence of LHCPs in plants, the amino acid next to arginine is tryptophan (W), which provided the ability to assay binding by fluorescence resonance energy transfer from tryptophan to Chl. Studies with 16-mer synthetic peptides containing the sequence —EIVHSRW— showed that indeed two molecules of Chl a bound to the peptide, and the binding was reduced by half when histidine was replaced with alanine (*42*). A similar replacement in the precursor of LHCP dramatically attenuated import of the protein into isolated chloroplasts (*43*). These studies led to the concept that binding of two molecules of Chl, one of which is Chl *b*, to the motif is an essential step in retention of LHCPs in the envelope during import and initiation of assembly of LHCII. Consequently, the generalized sequence, —ExxHxR—, was designated a 'retention motif' (*24*).

Expression of *Arabidopsis* Chl *a* Oxygenase in *Escherichia coil*

Synthesis of Chl *b* from Chl *a*, whether before or after esterification, is catalyzed by a membrane-bound activity (*44*), designated Chl *a* oxygenase (CAO) (*45-47*). An antiserum against this protein was obtained in rabbits after expression of *Arabidopsis CAO* cDNA in *E. coli*. A complex of proteins was immunoprecipitated from detergent-solubilized membranes from the first primary leaves of *Arabidopsis* seedlings. We examined this fraction by EPR spectroscopy to determine whether mononuclear iron and Rieske iron-sulfur centers, predicted by the amino acid sequence, were present. A signal indicative of the predicted high-spin mononuclear ferric iron was readily detected at g = 4.3 (Figure 5). A typical Rieske iron-sulfur complex was not detected, but a spectral feature was observed at g = 2.057. This signal is similar to that of bound Cu^{2+} (*48*), but adventitious transition metals should have been removed by a wash with 0.1 mM EDTA. Alternatively, the signal may indicate an unusual iron-sulfur complex. Most interestingly, a remarkably stable radical signal was detected at g = 2.0042, a value consistent with an unpaired electron on a tyrosine side-chain. Usually such radicals are stabilized by metal ions, particularly iron (*49*). To determine whether the EPR spectrum was indeed that of CAO, we

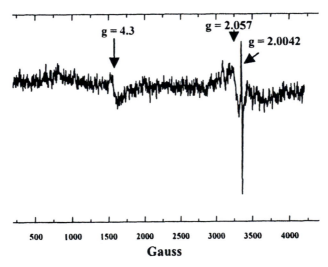

Figure 5. EPR spectrum, obtained at X-band (9.4 GHz) at a temperature of 7K, of the native CAO complex precipitated with protein A-Sepharose from solubilized Arabidopsis membranes incubated with antiserum against CAO. The sample was washed with 0.1 mM EDTA.

expressed the *CAO* cDNA in *E. coli*. The protein was recovered in the membrane fraction after centrifugation of broken cells. The EPR spectrum of the recombinant protein, shown in Figure 6 (*trace A*), was similar to that obtained for the native enzyme in the 'pull-down' preparation. Moreover, addition of Chl *a* to the recombinant enzyme quenched the radical signal (Figure 6, *trace B*).

The spectrum for CAO (Figure 5) was nearly identical to the EPR spectrum obtained by Jäger-Vottero et al. (*50*) of purified chloroplast envelope membranes. These investigators also detected the unusual g = 2.057 signal, which was not observed at temperatures higher than 20K, a characteristic of iron-sulfur centers. The similarities in these spectra suggested that the stable radical in the envelope membranes was contributed by CAO. It is unclear at this stage of the work whether the radical is involved in the reaction, possibly by abstraction of an electron or hydrogen atom from Chl *a*, analogous to the mechanism of action of lipooxygenase or ribonucleotide reductase, the latter of which also contains a stable tyrosine radical (*51,52*), or of cytochrome P450 mono-oxygenase (*53*). The redox potential of a tyrosine radical (+0.93 V) is sufficiently high to oxidize Chl *a* (E^{1} = +0.86 V, or near +0.5 V when paired with another polyaromatic molecule). A radical form of the substrate could then react with molecular oxygen to form a 7-hydroperoxy intermediate. The hydroperoxide could potentially be resolved to the hydroxymethyl derivative by a reductant such as mercaptoethanol, which would suggest a reaction analogous to that catalyzed by urate oxidase (*54*). Alternatively, the phenolic group of tyrosine may contribute an electron to formation of an activated iron-oxygen

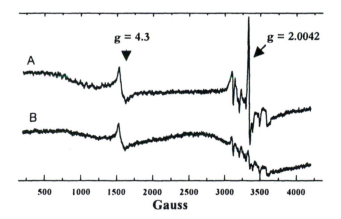

Figure 6. A, EPR spectra obtained at X-band (9.4 GHz) at 125K of a membrane pellet fraction of E. coli cells induced to express cDNA for the Arabidopsis Chl a oxygenase. To reduce recovery of the protein as inclusion bodies, cells were induced to synthesize the protein at 20°C. B, The sample as in A was incubated with Chl a prior to analysis by EPR. The signals between 3200 and 3700 Gauss were contributed by Mn in the samples, which were not washed with EDTA. The g = 2.057 signal was not observed at the higher temperature.

complex that acts as a mono-oxygenase to generate the 7-hydroxymethyl Chl intermediate, as described by Oster et al. (*55*). In their studies, they found an intermediate that eluted from a C_{18} HPLC column before Chl *b* in the position expected for 7-hydroxymethyl Chl. The mechanism they proposed involves two successive hydroxylation reactions to produce a *gem*-diol, which, as the hydrated form of an aldehyde, would be resolved by loss of water. However, 7-hydroxymethyl Chl is expected to be oxidized significantly less readily than Chl *a* itself, based on considerations of redox potentials of the substrates. These possible reaction pathways were proposed by Porra et al. (*56*), who determined that the oxygen in the 7-formyl group of Chl *b* is derived from molecular oxygen. Our studies provide an approach toward an analysis of the mechanism of this reaction.

Conclusions

Our evidence that Chl *b* synthesis is required to retain LHCPs in the chloroplast leads to two principal conclusions. Firstly, at least during the early stages of chloroplast development, the final stages of Chl synthesis, including that of Chl *b*, must occur in the envelope membranes, interaction of the proteins with Chl and assembly of a functional membrane must also be initiated in the envelope. Secondly, an understanding of the assembly of LHCs must include the chemistry of coordination bond formation between Chls and the proteins. These interactions are apparently necessary for correct folding of the protein, and they provide the specificity of occupancy of binding sites for Chl.

An understanding of the factors that enhance assembly and stability of light-harvesting complexes can possibly lead to engineering more robust organisms for a number of useful, 'green' processes. Algae provide a potential 'scrubber' of CO_2 in power plant effluents. Algae enriched in Chl *b* may be more stable under the lower pH environments in such effluents because of its greater stability to acidic pH values than Chl *a*, which results either because the Mg in Chl *b* repels protons or because it is hexacoordinated and thus sterically protected from displacement by protons. Conditions in photobioreactors that allow growth of algal cultures to achieve high cell densities gradually become light limited, and under these conditions production may become greater with organisms that have an expanded light-harvesting antennae.

It may also emerge, based on our data, that oxidation of Chl *a* to Chl *b* occurs by a radical mechanism, which would be sensitive to radical scavengers and anti-oxidants that are abundant in foods and nutritional supplements. Such chemicals in the environment could substantially deter growth of efficient, photosynthetic organisms. Much detailed work remains to be done on all aspects of the proposals made in this article.

Acknowledgements

We thank Dr. Judy Brusslan for providing the anti-CAO antiserum and Dr. Ayumi Tanaka for the gift of *GAO* cDNA.

References

1. Green, B.R.; Durnford, D.G. *Annu. Rev. Plant Physiol. Plant Mol. Biol.* **1996**, *47*, 685-714.
2. Durnford, D.G.; Deane, J.A.; Tan, S.; McFadden, G.I.; Gantt, E.; Green, B.R. *J. Mol. Evol.* **1999**, *48*, 59-68.
3. Hoober, J.K.; Eggink, L.L. *FEBS Lett.* **2001**, *489*, 1-3.
4. Eggink, L.L.; Park, H.; Hoober, J.K. *BMC Plant Biol.* **2001**, *1*, 2.
5. Phillips, J.N. In *Comprehensive Biochemistry;* Florkin, M.; Stotz, E.H., Eds.; Elsevier Scientific: Amsterdam, **1963**; Vol 9, pp 34-72.
6. Smith, K.M. In *Porphyrins and Metalloporphyrins;* Smith, K.M., Ed.; Flsevier Scientific: Amsterdam, 1975; pp 1-58.
7. Belanger, F.C.; Rebeiz, C.A. *Spectrochim. Acta* **1984**, *40A*, 807-827.
8. Rebeiz, C.A. In *Heme, Chlorophyll, and Bilins: Methods and Protocols:* Smith, A.G.; Witty, M., Eds.: Humana Press: Totowa, NJ, **2002**; pp 111-155.
9. Noy, D.; Yerushalmi, R.; Brumfeld, V.; Ashur, I.; Scheer, H.; Baldridge, K.K.; Scherz, A. *J. Am. Chem. Soc.* **2000**, *122*, 3937-3944.
10. Koerner, R; Wright, J.L.; Ding, X.D.; Nesset, M.J.M.; Aubrecht, K.; Watson, R.A.; Barber, R.A.; Mink, L.M.; Tipton, A.R.; Norvell, C.J.; Skidmore, K.; Simonis, U.; Walker, F.A. *Inorg. Chem.* **1998**, *37*, 733-745.
11. Dougherty, R.C.; Strain, N.H.; Svec, W.A.; Uphaus, R.A.; Katz, J.J. *J. Am. Chem. Soc.* **1970**, *92*, 2826-2833.
12. Ballschmiter, K.; Cotton, T.M.; Strain, N.H.; Katz J.J. *Biochim. Biophys. Acta* **1969**, *180*, 347-359.
13. Remelli, R.; Varotto, C.; Sandonà, à.D.; Croce, R.; Bassi, R. *J. Biol. Chem.* **1999**, *274*, 33510-33521.
14. Rogl, H.; Kühlbrandt, W. *Biochemistry* **1999**, *38*, 16214-16222.
15. Rogl, H.; Schödel, R.; Lokstein, H.; Kühlbrandt, W.; Schubert, A. *Biochemistry* **2002**, *41*, 2281-2287.
16. Reinsberg, D.; Booth, P.J.; Jegerschöld, C.; Khoo, B.J.; Paulsen, H. *Biochemistry* 2000, *39*, 14305-14313.
17. Tamiaki, H.; Yagai, S.; Miyatake, T. *Bioorg. Med. Chem.* **1998**, *6*, 2171-2178.
18. Paulsen, H.; Finkenzeller, B.; Kühlein, N. *Eur. J. Biochem.* **1993**, *215*, 809-816.
19. Kühlbrandt, W.; Wang, D.N.; Fujiyoshi,Y. *Nature* **1994**, *367*, 614-621.
20. Bassi, R.; Grace, R.; Cugini, D.; Sandonà, D. *Proc. Natl. Acad. Sci. USA* **1999**, *96*, 10056-10061.

21. Voigt, B.; Irrgang, K-D.; Ehlert, I.; Beenken, W.; Renger, G.; Leupold, D.; Lokstein, H. *Biochemistry* **2002**, *41,* 3049-3056.
22. Bossmann, B.; Grimme, L.H.; Knoetzel. J. *Planta* **1999**, *207,* 551-558.
23. Preiss, S.; Thornber, J.P. *Plant Physiol.* *1995*, *107,* 709-717.
24. Hoober, J.K.; Eggink, L.L. *Photosynth. Res.* **1999**, *61,* 197-215.
25. Park, H.; Hoober, J.K. *Physiol. Plant.* **1997**, *101,* 135-142
26. White, R.A.; Wolfe, G.R.; Komine, Y.; Hoober, J.K. *Photosynth. Res.* **1996**, *47,* 267-280.
27. Wolfe, G.R.; Park, H.; Sharp, W.P.; Hoober, J.K. *J. Phycol.* **1997**, *33,* 377-386.
28. Michel, H.P.; Tellenbach, M.; Boschetti, A. *Biochim. Biophys. Acta* **1983**, *725,* 417-424.
29. Chunaev, A.S.; Mirnaya, O.N.; Maslov, V.G.; Boschetti, A. *Photosynthetica* **1991**, *25,* 291-301.
30. Park, H.; Eggink, L.L.; Roberson, R.W.; Hoober, J.K. *J. Phycol.* **1999**, *35* 528-538.
31. Hoober, J.K.; Maloney, M.A.; Asbury, L.R.; Marks, D.B. *Plant Physiol.* **1990**, *92,* 419-426.
32. Maloney, M.A.; Hoober, J.K.; Marks, D.B. *Plant Physiol.* **1989**, *91,* 1100-1106.
33. Eggink, L.L.; Park, H.; Hoober, J.K. In *The Chloroplast: From Molecular Biology to Biotechnology;* Argyroudi-Akoyunoglou, J.H.; Senger, H., Eds.; Kluwer Academic Publishers: Dordrecht, the Netherlands, **1999**; pp 161-166.
34. Reinbothe, S.; Reinbothe, C. *Eur. J. Biochem.* **1996**, *237,* 323-343.
35. Timko, M.P. In *The Molecular Biology of Chloroplasts and Mitochondria in Chlamydomonas;* Rochaix, J-D.; Goldschmidt-Clermont, M.; Merchant, S., Eds.; Kluwer Academic Publishers: Dordrecht, the Netherlands, **1998**; pp 377-414.
36. White, R.A.; Hoober, J.K. *Plant Physiol.* **1994**, *106,* 583-590.
37. Hoober, J.K.; Argyroudi-Akoyunoglou, J. In *Chlorophyll Fluoresence: The Signature of Photosynthetic Efficiency, Advances in Photosynthesis and Respiration Series;* Papageorgiou, G.C.; Govindjee, Eds.; Kluwer Academic Publishers, Dordrecht, the Netherlands, **2004**, in press.
38. Kroll, D.; Meierhoff, K.; Bechtold, N.; Kinoshita, M.; Westphal, S.; Vothknecht, U.C.; Soll, J.; Westhoff, P. *Proc. Natl. Acad Sci. USA* **2001**, *98,* 4238-4242.
39. Westphal, S.; Heins, L.; Soll, J.; Vothknecht, U.C. *Proc. Natl. Acad. Sci. USA* **2001**, *98,* 4243-4248.
40. Dolganov, N.A.M.; Bhaya, D.; Grossman, A.R. *Proc. Natl. Acad. Sci. USA* **1995**, *92,* 636-640.
41. Funk, C.; Vermaas, W. *Biochemistry* **1999**, *38,* 9397-9404.
42. Eggink, L.L.; Hoober, J.K. *J. Biol. Chem.* **2000**, *275,* 9087-9090.
43. Kohorn, B.D. *Plant Physiol.* **1990**, *93,* 339-342.
44. Bednarik, D.P.; Hoober, J.K. *Science* **1985**, *230,* 450-453.

45. Tanaka, A.; Ito, H.; Tanaka, R.; Tanaka, N.K.; Yoshida, K.; Okada, K. *Proc. Natl. Acad. Sci. USA* **1998**, *95,* 12719-12723.
46. Espineda, G.E.; Linford, A.S.; Devine, D.; Brusslan, J.A. *Proc. Natl. Acad. Sci. USA* **1999**, *96,* 10507- 10511.
47. Tomitani, A.; Okada, K.; Miyashita, H.; Matthijs, H.C.P.; Ohno, T.; Tanaka, A. *Nature* **1999**, *400,* 159-162.
48. Roberts, A.G.; Bowman, M.K.; Kramer, D.M. *Biochemistry* **2002**, *41,* 4070-4079.
49. Stubbe, J; van der Donk, W.A. *Chem. Rev.* **1998**, *98,* 705-762.
50. Jäger-Vottero, P.; Dorne, A-J.; Jordanov, J.; Douce, R.; Joyard, J. *Proc. Natl. Acad Sci. USA* **1997**, *94,* 1597-l602.
51. Que, L. Jr.; Ho R.Y.N. *Chem. Rev.* **1996**, *96,* 2607-2624.
52. Yun, D.; Krebs, C.; Gupta, G.P.; Iwig, D.F.; Huynh, B.H.; Bollinger, J.M., Jr. *Biochemistry* **2002**, *41,* 981-990.
53. Hata, M.; Hirano, Y.; Hoshino, T.; Tsuda, M. *J. Am. Chem. Soc.* **2001**, *123,* 6410-6416.
54. Sarma, A.D.; Tipton, P.A. *J. Am. Chem. Soc.* **2000**, *122,* 11252-11253.
55. Oster, U.; Tanaka, R.; Tanaka, A.; Rüdiger, W. *Plant J.* **2000**, *21,* 305-310.
56. Porra, R.J.; Schäfer, W.; Cmiel, E.; Katheder, I.; Scheer, H. *Eur. J. Biochem.* **1994**, *219,* 671-679.

Agricultural Products
as Green Chemistry

Chapter 4

Plant Essential Oils as Green Pesticides for Pest and Disease Management

Murray B. Isman

Faculty of Agricultural Sciences, University of British Columbia, Vancouver, British Columbia V6T 1Z4, Canada

Certain plant essential oils show a broad spectrum of activity against pest insects and plant pathogenic fungi, and some oils have a long tradition of use in the protection of stored products. Recent investigations indicate that some chemical constituents of these oils interfere with the octopaminergic nervous system in insects. As this target site is not shared with mammals, most essential oil chemicals are relatively non-toxic to laboratory animals and fish in toxicological tests, and meet the criteria for "reduced risk" pesticides. Some of these oils and their constituent chemicals are widely used as flavoring agents in foods and beverages and are even exempt from pesticide registration in the United States. This special regulatory status combined with the wide availability of essential oils from the flavor and fragrance industries, has made it possible to fast-track commercialization of essential oil-based pesticides in the U.S.A. Though well received by consumers for use against home and garden pests, these "green pesticides" can also prove effective in agricultural situations, particularly for organic food production.

For the past five decades, synthetic insecticides have been the most important tool for pest management in agriculture, forestry, in human dwellings and in managed landscapes. In spite of ever-growing public concern for the environmental effects of pesticides (e.g. groundwater contamination, depletion of wildlife populations) and human health effects (e.g. chronic diseases such as cancer, developmental effects on children), and increasing calls for alternative means of pest management, it appears that synthetic insecticides will continue to remain the cornerstone of insect and plant disease control for at least another decade (1). Genetically modified (transgenic) crop plants expressing resistance factors such as the insecticidal protein from *Bacillus thuringiensis*, looked set to displace chemical pesticides from protection of the most important food and fiber crops as little as five years ago, but public distrust of the technology as applied to food production has slowed the adoption of this technology, insuring a place for pesticides.

Other alternatives to synthetic insecticides include microbial pesticides (such as *Bacillus thuringiensis* and insect baculoviruses), pheromones (used for mating disruption, trapping out and bait-and-kill strategies), and botanical insecticides. Botanicals such as pyrethrum, rotenone and nicotine were among the earliest insecticides used by farmers, with records dating back more than a century. Synthetic insecticides with greater potency, residual action and convenience of use largely displaced botanicals from the marketplace in the 1950s. In the case of plant pathogenic microorganisms, inocula are essentially everpresent (except in quarantined regions), and therefore prophylactic measures are normally required. There are some competitive microorganisms, but at this point alternatives to synthetic fungicides are largely limited to natural products with sufficient antifungal bioactivity to be efficacious in an agricultural context.

The Food Quality Protection Act (FQPA), enacted by the U.S. government in 1996, called for a reassessment of pesticide tolerances in food, with a particular view to cumulative risks (those from different chemicals but with a common mode-of-action) and aggregate risks (those from multiple sources of exposure), and with a special emphasis on safety to children. The U.S. Environmental Protection Agency had already in the early 1990s established the category of 'reduced-risk" pesticides. Reduced-risk products are those that (i) pose almost no risk to the user under proposed label uses and restrictions, (ii) break down rapidly and completely in the environment, (iii) leave little or no residues in food, and (iv) are compatible with integrated pest management (e.g. use of biocontrol agents). In this political 'environment', we might expect that pesticides based on plant natural products (i.e. botanicals) would have an opportunity to at least partially fill the void left by the many synthetic pesticides whose uses are or will be dramatically restricted or become deregistered under the provisions of the FQPA.

Many plant extracts or their active principles have demonstrated significant bioactivity against pest insects. Among these are several leads with potential for commercial development (2). Unfortunately, issues such as availability and regulatory concerns (product standardization, toxicity arising from microbial contamination) have curtailed commercial introduction of many potentially useful products. Derived from the seeds of the Indian neem tree, *Azadirachta indica* (Meliaceae), neem is one botanical that has recently been registered as a pesticide in North America and in Western Europe. As an insecticide, neem meets all the criteria of a reduced-risk product (unique mode-of-action, non-toxic to mammals, rapid degradation in the environment), but its commercial success in the U.S. has failed to live up to initial predictions mostly because of its high cost and limited availability (*3*).

Essential oils and their constituents as pesticides

Those materials obtained through steam distillation of plant material are referred to as "essential oils". They typically consist of highly complex mixtures of mono- and sesquiterpenoids and biogenetically-related phenols that confer unique aromas and flavours to plants (*4*). Some of these oils or their chemical constituents are well known as insect repellents and deterrents (e.g. oil of citronella)(*5*) and some have longstanding traditional uses as protectants of stored grain (*6,7*). In particular, aromatic plants from the families Lamiaceae and Myrtaceae have been well studied in this regard (Table I). The relatively high vapor pressures of essential oil constituents provides for both contact and fumigant action against a wide range of pests. Among the more effective compounds are eugenol, from cloves (*Eugenia caryophyllus*, Myrtaceae), thymol from garden thyme (*Thymus vulgaris*, Lamiaceae) and cinnamaldehyde from cinnamon (*Cinnamomum zeylandicum*, Lauraceae)(*8*)(Figure 1). Some essential oils have also been demonstrated to have fumigant action against plant pests (aphids and spider mites) in greenhouse trials (*9*) and against subterranean termites, *Coptotermes formosansus* (*10*).

Table 1. Plant essential oils with demonstrated pesticidal actions in insects and/or fungi.

Common name	Species (Family)	Major constituent(s)
Cloves	*Eugenia caryophyllus* (Myrtaceae)	eugenol
Eucalyptus	*Eucalyptus globulus* (Myrtaceae)	1,8-cineole (eucalyptol)
Lemon grass	*Cymbopogon nardus* (Poaceae)	citronellal, citral
Pennyroyal	*Mentha pulegium* (Lamiaceae)	(+)-pulegone
Thyme	*Thymus vulgaris* (Lamiaceae)	thymol, carvicrol
Rosemary	*Rosmarinus officinalis* (Lamiaceae)	1,8-cineole, α-pinene
Cinnamon	*Cinnamomum zeylandicum* (Lauraceae)	cinnamaldehyde

SOURCE: Reproduced with permission from Reference 27. Copyright 1999 Royal Society of Chemistry

44

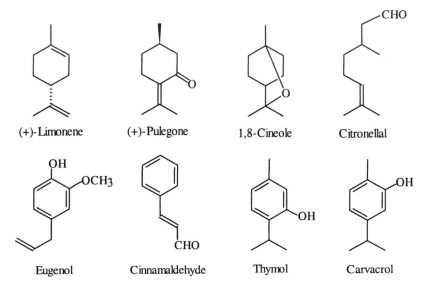

(+)-Limonene   (+)-Pulegone   1,8-Cineole   Citronellal

Eugenol   Cinnamaldehyde   Thymol   Carvacrol

Figure 1. Some major essential oil constituents with insecticidal and/or fungicidal activities. Top row: monoterpenes; bottom row: phenols.

Certain essential oil constituents are very effective against domestic pests such as cockroaches, *Periplaneta Americana* and *Blattella germanica*, and houseflies, *Musca domestica* (*11,12*). Toxicity to several agricultural pests, e.g. western corn rootworm, *Diabrotica virgifera* and European cornborer, *Ostrinia nubilalis* (*13*), tobacco cutworm, *Spodoptera litura* and green peach aphid, *Myzus persicae* (*4*), confirm that these chemicals are generally active against a broad spectrum of pests. In addition, selected essential oils and constituents have proven effective against the *Varroa* mite, a serious ectoparasite of the honey bee (*14*). Figure 2 shows the comparative toxicity of a number of common plant essential oils to 3rd instar larvae of the cabbage looper (*Trichoplusia ni*) following topical administration.

Although several essential oil monoterpenoids and phenols have demonstrated toxicity to one or more pests, extensive investigations into structure-activity relations have not provided much insight nor predictive power (*15,16*). Overall, the toxicity of any particular essential oil compound to insects can best be described as idiosyncratic. For example, in comparing toxicity of two phenols (eugenol and carvacrol) and two monoterpenes (α-terpineol and terpinen-4-ol) to five species of insects and one mite species, carvacrol was clearly most toxic to one species, eugenol most toxic to another, terpinen-4-ol to two others, eugenol and carvacrol equitoxic to the fifth, and eugenol and terpinen-4-ol equitoxic to the sixth. On the other hand, we have demonstrated that certain combinations of essential oil compounds, particularly those including *trans*-anethole, are synergistic with respect to their toxicity to the tobacco cutworm *Spodoptera litura* (*17*). Thus there is the opportunity to achieve a

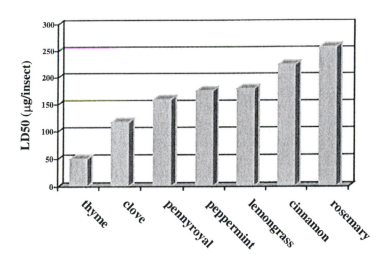

Figure 2. Toxicity of some common plant essential oils to the cabbage looper.

desired spectrum of activity and optimal efficacy through blending of specific constituents or oils.

The rapid knockdown of insects by certain essential oils or their constituents is indicative of a neurotoxic mode-of-action. Certain mononterpenoids competitively inhibit acetylcholine esterase *in vitro* (*18,19*), but this action does not appear to correlate with overall toxicity to insects *in vivo*. Experimental evidence now points to the octopaminergic nervous system of the insect as the likely target site, including inhibition of ^{3}H-octopamine binding in a cockroach nerve cord preparation in the presence of certain essential oil compounds (*20*). Formamidine insecticides (e.g. chlordimeform) target the octopaminergic nervous system in insects, and studies with these chemicals over 20 years ago revealed interesting and potentially important behavioral and physiological effects at sublethal doses (*21*). Behavioral effects of essential oils and their constituents, e.g feeding deterrence, repellency, are consistent with this mode-of-action (*4,5,17*).

The absence of octopamine receptors in vertebrates at least in part accounts for the relative low mammalian toxicity of essential oils or their monoterpenoid constituents. As such, pesticides based on plant essential oils have excellent mammalian selectivity. A handful of these compounds, in purity, have rat oral acute LD_{50} values in the 2-3 g kg^{-1} range, but commercial insecticides consisting of proprietary mixtures of essential oil compounds produce less than 50% mortality in rats (and often no mortality) at 5 g kg^{-1} (E. Enan, unpubl. data), the upper limit required for acute toxicity tests by most pesticide regulatory agencies. Some of the essential oils and their constituents are commonly used as culinary herbs and spices and are even exempt from toxicity data requirements in the United States.

Environmental impacts of essential oil-based pesticides can also be expected to be minimal or at the least more favourable compared to conventional synthetic pesticides. Static water tests with juvenile rainbow trout (*Oncorhynchus mykiss*), indicated that thymol is about 400 times less toxic than the botanical insecticide pyrethrum, and 4,000 times less toxic than the organophosphate insecticide azinphosmethyl (*22*).

Given their broad spectrum neurotoxicity, selectivity favoring natural enemies and biocontrol agents should not be expected. In laboratory contact toxicity bioassays using a precommercial formulated essential oil insecticide, the LD_{50} for 3rd instar green lacewing larvae (*Chrysoperla carnea*) was only about 20% higher than that of 3rd instar larvae of the obliquebanded leafroller (*Choristoneura rosaceana*), a fruit and berry pest (D. MacArthur and M. Isman, unpubl. data). We have also found that predatory mites (*Phytoseilus persimilis*) are as susceptible as the twospotted spider mite (*Tetraychus urticae*).

On the other hand, most monoterpenoids degrade rapidly and completely in soil and water (half-lives < 48 hr) and are rapidly lost from plant foliage through volatilization (J. Coats, R. Bradbury, M. Isman, unpubl. data). The lack of persistence on foliage favors immigration or reintroduction of biological control agents, and thus these pesticides are compatible with integrated pest management. They clearly meet the criteria for 'reduced-risk' pesticides.

Commercialization of essential oil-based and other botanical insecticides

A substantial worldwide research effort aimed at the discovery and development of new botanical insecticides with acceptable efficacy and minimal health and environmental impacts, has unfortunately resulted in very few products actually reaching the marketplace. Although a number of plant preparations meet the criteria for efficacy (e.g. seed extracts of *Annona*, *Citrus* and *Piper* species)(2), registration (including the cost of toxicological evaluations) remains a formidable barrier to commercialization for many prospective natural pesticides. Other barriers to commercialization (23) are (i) availability of the starting material on a sustainable and consistent basis, (ii) the need for chemical standardization and quality control, and (iii) costs of the raw materials and their refinement.

Neem insecticides certainly meet the criteria of efficacy and safety, but their commercial success in North America and Europe has been impeded by supply problems, product inconsistency and high product costs. Pyrethrum, the world leader among botanical insecticides, has suffered in recent years from supply problems and it too is considered relatively expensive (though less so than neem)(Table II).

Exemption of certain essential oils as pesticide active ingredients from registration in the U.S. in 1996 (Environmental Protection Agency, 40 CFR Part 152) has allowed some American companies to bring essential oil-based pesticides to market more rapidly than would normally be required for a conventional product. Exploitation of essential oils or their constituents for pesticide production is also facilitated by the worldwide production and marketing of these chemicals as flavoring agents and fragrances; they are widely available and some are relatively inexpensive. Mycotech Corporation produces Cinnamite[TM], an insecticide/miticide/fungicide for use in greenhouse crops, and Valero[TM], a miticide/fungicide for use on grapes, berries and tree fruits. Cinnamon oil, containing 30% cinnnamaldehyde, is the active ingredient in both products.

EcoSMART Technologies holds U.S. and international patents on the use of monoterpenoids as octopamine blockers for insect control and compositional patents on several novel products, and has introduced more than two dozen pesticides in the past three years. Among these are aerosol, dust and other formulations containing proprietary/patented mixtures of essential oils compounds for control of domestic pests (viz. cockroaches, ants, fleas, flies).

Table II. Comparison of botanical insecticides in North America

Pesticide:	Pyrethrum	Neem	Essential Oils
Major sources	Kenya (Australia*)	India (Australia*)	Worldwide
Availability	Limited in recent years	Limited at present	Widely available
Approx. price (USD/kg)	45-60	125-200	10-25
% actives (technical grade)	20	10	>90
Major use	Insecticide	Insecticide	Flavors and fragrances

*pyrethrum production in Tasmania is expected to begin in 2002 or 2003; neem production in North Queensland is anticipated around 2005

These are marketed in concentrated form for pest control professionals under the brand name EcoPCO™, and in less concentrated form to the general consumer through the Bioganic™ brand. Most recently, agricultural insecticide/miticides have been released under the names Hexacide™ and Ecotrol™. These latter products, containing rosemary oil as the active ingredient, have proven efficacious in field trials against aphids (*Myzus persicae*, *Aphis gossypii*, *Chaetosiphon fragaefolii*), whiteflies (*Trialurodes vaporariorum*), thrips (*Frankliniella occidentalis*) and mites (*Tetranychus urticae*, *T. pacificus*) on a variety of crops including cotton, strawberry, grapes, squash and flowering ornamentals. We have also demonstrated reductions in populations of lepidopteran pests on cabbage in the field following treatment with an essential oil-based insecticide, comparable to those seen with pyrethrum.

To what extent control of insect and mites pests in the field arises from behavioral effects (i.e. deterrence, repellency), rather than contact toxicity, remains to be determined. For example, rosemary oil has been shown to repel the green peach aphid (*Myzus persicae*) in the laboratory and inhibit settling of aphids under screenhouse conditions (*24*). It is important to note that the essential oil products are less potent than most other pesticides, so greater quantities are required. However the environmental non-persistence of the oils makes this fact acceptable. Another important consideration is phytotoxicity. Like mineral oils and other plant oils, some essential oils can be phytotoxic to certain sensitive plants at rates not far above that required for insect/disease control. Before commercial introduction, sufficient field testing is required to determine that the product can be used safely on plant foliage. A final word of caution – as natural products, plant essential oils are subject to considerable genetic, temporal and geographical variation in their chemistry and biological activity. For example, we recently tested ten samples of rosemary oil (representing five producer/suppliers) and found that LD_{50} values for these oils applied topically to 3rd instar armyworms (*Pseudaletia unipuncta*) ranged from 167 to 372 µg per insect (M. Isman, unpubl. data).

As the Mycotech products suggest, certain essential oil constituents that are insecticidal and miticidal are effective against plant pathogenic fungi as well. Investigations of essential oils against fungi actually predate those against insects. Screening studies indicate that some essential oils and their constituents are effective against a broad spectrum of plant pathogenic fungi (25,26). With respect to fungi, the phenols (thymol, carvacrol, eugenol) appear more consistently effective than the monoterpenes. In my laboratory we have shown the efficacy of these compounds *in vitro* against a range of plant pathogens including *Verticillium* and *Fusarium* species (Isman, unpubl. data). EcoSMART Technologies have developed a rosemary oil-based fungicide for agricultural use (Sporan[TM]). Preliminary field trials to date have shown that this product can be as effective as conventional fungicides in protecting grapes from powdery mildew (*Uncinula necator*) and against brown patch (*Rhizoctonia solani*) on turfgrass.

Summary

Certain plant essential oils and/or their chemical constituents, applied singly or in specific combinations, are effective against a range of important arthropod pests and plant pathogens. With few exceptions they are relatively non-toxic to mammals and non-persistent in the environment (27). Their long history of use as flavorings for foods and beverages is further indication of their safety to humans.

The regulatory situation in the United States currently favors introduction of "reduced-risk" pesticides, and essential oil-based pesticides fit these criteria well. Essential oil-based pesticides are not a panacea for pest and disease control, but as conventional products leave the marketplace, growers will have fewer choices in their arsenal. Essential oil-based pesticides and other botanicals such as neem should find particular acceptance in market niches where a premium is placed on worker safety and environmental protection, for example in organic food production.

Acknowledgments

Based on a paper presented at the International Symposium, *"Development of Natural Pesticides From Forest Resources"*, Seoul, Korea, October 2001. Research in the author's laboratory is funded by grants from the Natural Sciences and Engineering Research Council (Canada) and EcoSMART Technologies Inc. Dr. Essam Enan and David Lindsay kindly read the manuscript prior to communication.

References

1. Benbrook CM. *Pest Management at the Crossroads.* Consumers Union, Yonkers, NY, 1996.
2. Isman, M.B. *Rev. Pestic. Toxicol.* **1995,** *3*, 1-20.
3. Isman, M.B. In: *Neem: Today and in the New Millenium;* Koul, O.; Wahab, S., Eds.; Kluwer, Dordrecht, The Netherlands, *in press.*
4. Isman, M.B. Plant essential oils for pest and disease management. *Crop Protection* **2000,** *19,* 603-608.
5. Peterson, C.J.; Coats, J.R. *Pesticide Outlook* **2001,** *12,* 154-158.
6. Shaaya, E.; Ravid, U.; Paster, N.; Juven, B.; Zisman, U.; Pissarev, V. *J. Chem. Ecol.* **1991,** *17*, 499-504.
7. Regnault-Roger, C.; Hamraoui, A.; Holeman, M.; Theron, E.; Pinel, R. *J. Chem. Ecol.* **1991,** *19*, 1233-1244.
8. Huang, Y.; Ho, S.H. *J. Stored Prod. Res.* **1998,** *34*, 11-17.
9. Tunc, I.; Sahinkaya, S. *Entomol. exp. appl.* **1998,** *86,* 183-187.
10. Cornelius, M.L.; Grace, J.K.; Yates III, J.R. *J. Econ. Entomol.* **1997,** *90,* 320-325.
11. Coats, J.R.; Karr, L.L.; Drewes, C.D. *Amer. Chem. Soc. Symp. Ser.* **1991,** *449,* 306-316.
12. Ngoh, S.P.; Hoo, L.; Pang, F.Y.; Huang, Y.; Kini, M.R.; Ho, S.H *Pestic. Sci.* **1998,** *54,* 261-268.
13. Lee, S.; Tsao, R.; Peterson, C.J.; Coats, J.R. *J. Econ. Entomol.* **1997,** *90,* 883-892.
14. Calderone, N.W.; Twilson, W.; Spivak, M. *J. Econ. Entomol.* **1997,** *90,* 1080-1086.
15. Rice, P.J.; Coats, J.R. *Pestic. Sci.* **1994,** *41*, 195-202.

16. Tsao, R.; Lee, S.; Rice, P.J.; Jensen, C.; Coats, J.R. *Amer. Chem. Soc. Symp. Ser.* **1995,** *584,* 312-324.
17. Hummelbrunner, L.A.; Isman, M.B. *J. Agric. Food Chem.* **2001,** *49,* 715-720.
18. Miyazawa, M.; Watanabe, H.; Kameoka, H. *J. Agric. Food Chem.* **1997,** *45,* 677-679.
19. Lee, S.-E.; Lee, B.-H.; Choi, W.-S.; Park, B.-S.; Kim, J.-G.; Campbell, B.C. *Pest Management Sci.* **2001,** *57,* 548-553.
20. Enan, E. *Comp. Biochem. Physiol. C* **2001,** *130,* 325-337.
21. Hollingworth, R.M.; Lund, A.E. In *Insecticide Mode of Action'* Coats, J.R., Ed. Academic Press, New York, 1982; pp 189-227.
22. Stroh, J.; Wan, M.T.; Isman, M.B.; Moul, D.J. *Bull. Environ. Contam. Toxicol.* **1998,** *60,* 923-930.
23. Isman, M.B. *Phytoparasitica* **1997,** *25,* 339-344.
24. Hori, M. *J. Chem. Ecol.* **1998,** *24,* 1425-1432.
25. Muller-Riebau, F.; Berger, B.; Yegen, O. *J. Agric. Food Chem.* **1995,** *43,* 2262-2266.
26. Wilson, C.L.; Solar, J.M.; El Ghaouth, A.; Wisniewski, M.E. *Plant Disease* **1997,** *81,* 204-210.
27. Isman, M.B. *Pesticide Outlook* **1999,** *10,* 68-72.

Chapter 5

Adhesive Produced by Microorganisms

J. Combie[1], A. Haag[2], P. Suci[2], and G. Geesey[2]

[1]Montana Biotech, 1910–107 Lavington Court, Rock Hill, SC 29732
[2]Montana State University, 123 Huffman Building, Bozeman, MT 59717

A water-based adhesive from non-petrochemical feedstock was produced by an efficient microbial fermentation process. The adhesive was susceptible to water but very resistant to solvents such as jet fuel. After curing, the adhesive could be re-moistened and used again to bond surfaces together. Shear strength on anodized aluminum averaged 819 psi and tensile strength ranged from 500 to 1500 psi depending on the substrate and production method. The parent adhesive was modified to several water resistant forms which maintained good adhesive strength.

Introduction

Adhesives often include toxic volatile organic compounds (VOCs) such as toluene, methylethylketone, and xylene. In sunlight, VOCs and nitrogen oxides produce ozone. Other VOC interactions contribute to the formation of photochemical smog. At elevated levels, VOCs present both a health and fire hazard and degrade the environment. The US used 5.6 billion pounds of adhesives in 2001. Sixteen percent of these were solvent based systems (1), releasing several hundred million pounds of VOCs into the atmosphere.

The search for environmentally compatible adhesives has turned to such examples of nature as the tenacious adherence of barnacles and mussels (2). Although the properties have indeed been spectacular, production of these adhesives on a commercial scale has been problematic.

53

It is well known that many microorganisms produce extracellular polymeric substances (EPS) with adhesive properties (*3,4,5,6*). The advantage of using microorganisms over mollusks is that the technology for commercial scale production of microorganisms is well established. The work reported here focuses on use of readily-grown microorganisms as a source of novel, environmentally friendly adhesives at a reasonable cost.

Experimental

Screening of Microorganisms

350 isolates from an in-house culture collection were raised in liquid culture and screened for those producing adhesive EPS. Adhesives were ranked by tensile strength on bare aluminum.

Optimized Production Process

Microorganisms were grown in the following liquid medium (gm/l): 0.08 $K_2HPO_4 \cdot 3H_2O$, 0.1 $MgSO_4 \cdot 7H_2O$, 0.001 $FeSO_4 \cdot 7H_2O$, 21 sucrose, 0.2 $(NH_4)_2SO_4$, 2.0 citric acid. The pH of the medium was adjusted to 6.4 with 10 N NaOH prior to sterilization. Following 2 days of growth at 35 °C, spent medium was mixed with -20 °C solvent such as isopropyl alcohol, causing the adhesive to precipitate out of solution. The mixture was centrifuged at 1500 g for 10 min, the supernatant was discarded and the putty-like adhesive collected from the bottom of the container.

Adhesion Testing Procedure

Tensile Strength

For determination of tensile strength, a pair of 2024 bare aluminum cylindrical adherends with one quarter square inch surface area and 1 inch long, were joined with the test adhesive. The two cylindrical adherends were held together in a constant force fixture for 1 hour, to allow the adhesive to set up. Curing was by drying so the cure time varied with the precipitating solvent and the adhesive concentration.

After curing, the adherends were pulled apart using a manually operated Mark 10, Model BGI, a force measuring instrument. The tensile strength was generally determined on 5 pairs of adherends and the average recorded.

For solvent resistance determination, after curing the adherends were submerged in the solvent to be tested for 48 hours. The adherends were then pulled apart as described above.

Shear Strength

2024 aluminum T3 coupons, 1 inch wide and 4 inches long, were coated at Barry Avenue Plating, Los Angeles, CA. Coupons were divided into 3 groups as follows: (1) Anodized, (2) Anodized and epoxy coated, and (3) Anodized, epoxy undercoat and CARC topcoat Specifications were as follows:

- Anodized: AMS 2471 sulfuric acid. This specification includes a hot water seal which seals the aluminum but makes a poor surface for adhesives. Through a misunderstanding, this hot water seal was included. Therefore, adhesion to these surfaces would be expected to be significantly less as compared with alternative anodization methods.
- Epoxy: MIL-P-23377
- CARC (Chemical Agent Resistant Coating): MIL-C-53039

Particular attention was paid to cleaning coupon surfaces ensuring the surface was fully wetted by the adhesive and completely covered the test area. Surfaces were cleaned with ethanol or EnSolv (n-propyl bromide, Enviro Tech, Melrose Park, IL). Adhesive was applied to a 1 inch by 0.5 inch area of one coupon and the second coupon overlapped that area. Coupons were clamped together for 1 hour to allow the adhesive to set up.

Shear strength testing was done on an Instron Universal Testing Instrument (Screw Drive), Model Number TTC with an Instron Tensile Load Cell, Model D, 1000 pound full scale range. ASTM Specification D1002-99 was followed. All shear strength testing was performed at Redstone Arsenal by Alexander Steel.

Water Resistant Derivatives

A series of methyl ether and acetate ester derivatives of varying degrees of substitution were prepared. Methylation was performed under anhydrous conditions using dimsyl potassium base in dimethylsulfoxide and methyl iodide. Acetylations were performed with acetic anhydride in pyridine which contained a small amount of water to aid solubilization of the starting material. Products were isolated either by precipitation into water (higher degrees of substitution), or dilution with water and dialysis if water soluble (lower degrees of substitution). Incorporation of methyl ether and acetate ester groups were confirmed by infrared and nuclear magnetic resonance spectroscopy.

The water solubility of the products ranged from completely soluble to a highly swollen gel to completely insoluble (but organic soluble). Each was screened for adhesive strength on aluminum and if favorable, the materials were then tested for water resistance after curing by soaking in water at 20 °C and by storage in a 75% humidity chamber at ambient temperature.

Results and Discussion

Screening of Microorganisms

A single organism was selected as the focus of the majority of the work reported on here, based on results of flatwise tension tests. The organism has not yet been identified. It should be noted that organisms used by others looking for microbial adhesives such as *Pseudomonas fluorescens* (*4*), *P. fragi* (*7*), and *P. putida* (*8*) produced no adhesive under our conditions.

Production Process

Medium

Numerous experiments were conducted to optimize the composition of the growth medium. Twelve different media with up to eight modifications of each medium were tested. The organism grew on most of the media and in only a few cases, was no adhesive produced. Adhesive production did require the presence of sucrose; three sources of molasses, glucose, fructose and maltose could not be substituted. All carbon sources tested permitted growth of the microorganism but use of any carbon source other than sucrose resulted in recovery of no adhesive. Only a few nitrogen sources were tested, but it was found that substitution of nitrate for ammonium ion resulted in absence of recoverable amounts of adhesive. Removal of iron decreased the yield of adhesive by about half, but amounts ranging from 0.001 to 0.020 gm/l 0.001 $FeSO_4 \cdot 7H_2O$ appeared to have little impact on the production. When magnesium was absent, the microorganism did not grow, but all levels tested between 0.1 and 0.8 gm/l $MgSO_4 \cdot 7H_2O$ had little impact on adhesive production.

It is known that stress can increase EPS production. Therefore several stress factors were introduced into the growth medium. Although phosphate was needed, the less phosphate available (down to 0.01 gm/l $K_2HPO_4 \cdot 3H_2O$), the greater the yield of adhesive. Since this was a poorly buffered medium, it was found that 0.04-0.08 gm/l $K_2HPO_4 \cdot 3H_2O$ resulted in more reliable production.

Another set of experiments was done to compare the effect of placing screens made out of autoclavable plastic, aluminum or copper in the fermenters. It was anticipated that the stress of high copper levels might increase the polymer production. It was also theorized that providing greater surface area might also increase the amount of polymer produced. No significant effects were seen and this line of the investigation was dropped.

Precipitating solvent

Since the adhesive is soluble in the aqueous medium, some method for recovering the adhesive had to be employed. Both solvents and ammonium sulfate were tested. All solvents tested that were miscible with water, were satisfactory. Although minor differences were seen in the yield, solvent cost and the ease of recycling would need to be considered before a choice could be made. Satisfactory precipitating solvents were methanol, ethanol, isopropyl alcohol, acetone and ethyl lactate (Table I).

Table I. Effect of Precipitating Solvent on Adhesive Yield and Tensile Strength

Solvent	Yield gm/l	Tensile Strength psi
Methanol	6.47	1346
Ethanol (denatured)	6.87	1136
Ethanol (not denatured)	7.47	978
Isopropyl alcohol (reagent)	7.54	1640
Isopropyl alcohol (certified)	7.60	1005
Acetone	7.34	>1795
Ethyl lactate	9.14	1427

NOTE: Tensile strength is average of 5 replicates. One part of spent medium was mixed with two parts of solvent. Yield is wet weight.

Use of precipitating solvents that were not miscible with water did not result in recovery of adhesive. Those tested in this category were ethyl acetate, hexane, n-butanol and sec-butanol. Precipitation with 10, 40 and 70% ammonium sulfate was also unsuccessful.

Several experiments were done to determine the optimum ratio of the precipitating solvent to the spent medium. Ratios tested ranged from 1 part spent medium to one-eighth part solvent through 1 part spent medium to eight parts solvent. Although the optimum varied slightly with the solvent, the optimum was generally in the range of 1:1 to 1:2 spent medium to solvent.

Appearance

The adhesive, in both parent and derivative forms, is white and sticky. The viscosity is similar to putty although this varies with the ratio and specific solvent used in the precipitation. Before application, the viscosity was adjusted. Usually mixing with equal parts of water, provided a readily spreadable adhesive. For some applications, one part of adhesive was mixed with up to 3 parts of water.

The adhesive can be dried to a fluffy, white powder by either freeze drying or simple evaporation. It can be reconstituted with water to the desired viscosity.

Chemical Analysis and Molecular Weight

Molecular weight analysis was done by size exclusion chromatography. Generally, molecular weights were in excess of 1 million.

Classical colorimetric analysis indicated the dried parent adhesive is 95% carbohydrate, 1-3% protein and 1% uronic acid.

Adhesion Testing

Curing

Although the adhesive sets up sufficiently for coupons to be handled within one hour at ambient temperature, full curing requires the adhesive to be well dried. Since this is a water based adhesive, drying requires extended periods of time. In an effort to decrease the drying time, a series of experiments were done where the glued coupons were incubated at increased temperatures for various time periods. The standard procedure now is to incubate glued coupons at 35 °C for 7 days. The drying chamber averages 23% humidity at this temperature.

Adhesive Reversibility

The adhesive could be cured, the cylindrical adherends pulled apart, the surface moistened, the cylindrical adherends stuck together and cured again and then pulled apart the second time with no loss in adhesive strength. The curing mechanism appears to be drying.

Resistance to Solvents

One of the more interesting properties of the adhesive was its resistance to the effects of various solvents. Cylindrical adherends were glued with the parent adhesive. Following a 10 day cure at ambient temperature, they were soaked in solvent for 48 hours, dried for 1 hour and tensile strength measured (Table II).

Heat Stability

It was found that the microbial adhesive could withstand high temperatures, at least for a short period of time. In one experiment, after samples had been glued and cured, they were autoclaved (121 °C, 15 psi, 25 minutes). They were

Table II. Effect of Solvent on Cured Adhesive

Solvent	Tensile Strength psi
Jet A fuel	932
d-limonene	922
Methylethylketone	805
Toluene	766
Control - air	728

NOTE: Tensile strength is average of 10 replicates.

Table III. Tensile Strength Not Decreased by Autoclaving

Sample	Tensile Strength psi
Autoclaved	811
Control (not autoclaved)	730

NOTE: Tensile strength is average of 5 replicates.

again allowed to cure (moisture in the autoclave disrupted the bonding), and it was found they retained the original adhesive strength (Table III).

Comparison with Other Natural Polymers

There are a number of other natural polymers with adhesive properties. Some are of microbial origin, while others are derived from plants. Table IV shows a comparison of tensile strength between the microbial adhesive reported on here and other natural adhesives.

Water resistant derivatives

The parent adhesive was not particularly resistant to water and lost 80% of its strength after 30 min water immersion (Table V). A similar effect was observed after incubation in a 75% relative humidity chamber at ambient temperature.

Table IV. Comparison with Other Natural Polymers

Polymer	% Solids	Tensile Strength psi
Montana Biotech parent adhesive	25	991
Cooked corn starch	25	691
Dextran	25	479
Carboxymethylcellulose	17	193
Guar gum	8	63
Xanthan gum	17	33
Sodium alginate	17	20

NOTE: Tensile strength is average of 9 or 10 replicates.

Table V. Tensile Strength and Effect of Moisture on Parent Adhesive and Derivatives

	Dry	0.5 hr Soak	2 hr Soak	4 hr Soak	7 Days 75%
Me - A	1546	228	374	260	
Me - B	>793		30	0	138
Ac - C	874	575	324		674
Ac - D	630	247		18	289
Parent -E	809	114	22	40	
Parent - F	829				91

NOTE: Dry refers to coupons after 7 day curing at 35 °C. Soak refers to coupons being cured for 7 days, then soaked in ambient temperature water and 75% refers to coupons being cured for 7 days, then held in a 75% humidity chamber. Me A and B are 2 different methylated derivative samples, Ac C and D are 2 different acetylated derivative samples and Parent E and F are 2 different samples of the parent adhesive. Tensile strength is average of 5 replicates.

Samples with greater water resistance were prepared. The polar hydroxyl groups in the parent adhesive promote adhesion to polar surfaces like aluminum but they are also hydrophilic and lead to a low water resistance. The hydrophilicity of the adhesive may be reduced by derivatization of the hydroxyl groups with less polar groups like ethers and esters. The degree of substitution (ratio of derivatized to underivatized hydroxyl groups) can be controlled to maximize the water resistance without sacrificing adhesive strength. As expected, the derivatives produced a range of adhesive strength and water resistance and the two generally varied inversely as the degree of substitution was changed.

Substrate Effects

For shear testing, the microbial adhesive was tested on 3 different surfaces, anodized, epoxy, and CARC. Table VI shows a summary of this data.

Table VI. Shear Strength (psi) of Adhesive on Coated Aluminum Coupons

	Parent	OMe derivative	3M 4799
Anodized	819	930	171
Epoxy	576	541	146
CARC	651	533	185

NOTE: Data points are averages of 5 replicates. 3M 4799 is a commercially available adhesive included here for comparison.

Toxicology

Two toxicology tests were performed on the parent compound. It was found to be non-cytotoxic, meeting the requirements of the Agar Diffusion Test, ISO 10993. Endotoxin was found present in the preparation. Dilutions were done out to 1:256 and the corrected Endotoxin Units/ml was reported as greater than 615. By way of comparison, xanthan gum, a commercially available microbially produced polysaccharide used as a food additive, was also tested and found to have greater than 1280 Endotoxin Units/ml.

Conclusions

An adhesive was produced from a non-petrochemical feedstock by an efficient microbial fermentation. Made from a renewable resource, this water based adhesive is environmentally friendly. Its resistance to solvents makes it particularly useful in certain industrial settings. On anodized aluminum, shear strength averages 819 psi. Improved water resistance can be obtained with derivatization with ether or ester groups.

Acknowledgements

The authors gratefully acknowledge the support of the Strategic Environmental Research and Development Program (SERDP) and the Montana Board of Research and Commercialization Technology for funding the work reported here.

References

1. Nick, D. P. The North American Adhesive and Sealant Industry Convencion Annual Secciones. **2002**, URL http://www.ascouncil.org/news/ppt/dpna_asc_may2002.ppt
2. Waite, J. H.; Tanzer, M. L. Science. **1981**, *212*, 1038-1040.
3. Allison, D. G.; Sutherland I. W. J. Gen. Microbiol. **1987**, *133*, 1319-1327.
4. Pringle, J. H.; Fletcher, M. J. Gen. Microbiol. **1986**, *132*, 743-749.
5. Suci, P. A.; Frolund, B.; Quintero, E. J.; Weiner, R. M.; Geesey, G. G. Biofouling. **1995**, *9*, 95-114.
6. Whitfield, C. Can. J. Microbiol. **1988**, *34*, 415-420.
7. Zottola, E. A. Biofouling. **1991**, *5*, 37-55.
8. Read, R. R.; Costerton, J. W. Can. J. Microbiol. **1987**, *33*, 1080-1090.

Chapter 6

Enhancement of Ethanol Yield from the Corn Dry Grind Process by Conversion of the Kernel Fiber Fraction

B. S. Dien[*], N. N. Nichols, P. J. O'Bryan, L. B. Iten,
and R. J. Bothast

Fermentation Biotechnology Research Unit, National Center
for Agricultural Utilization Research, Agricultural Research Service,
U.S. Department of Agriculture, 1815 North University Street,
Peoria, IL 61604
[*]Corresponding author: telephone: 309–681–6270; fax: 309–681–6427;
email: dienb@ncaur.usda.gov

Approximately 50% of the corn processed for fuel ethanol in the U.S. is dry grinded. Dry grinding yields 2.7-2.8 gal of ethanol per bushel of corn. This ethanol yield could be increased if the fiber component of the corn kernel was also converted into ethanol. Currently, the kernel fiber in a dry grind plant is collected after the fermentation in a solids cake by centrifugation and termed Distillers Wet Grains (DWG). Three different samples of DWG were analyzed and found to contain 14.7- 18.1% glucans and 34.9-40.5% total carbohydrates. We have successfully converted the fiber component of DWG into ethanol using either an industrial *Saccharomyces cerevisiae* strain (Y-2034) or ethanologenic *Escherichia coli* strain FBR5. For the *S. cerevisiae* fermentation, DWG was pretreated with dilute acid and simultaneously saccharified and fermented (SSF) by adding cellulase, beta-glucosidase, and glucoamylase along with the yeast. The ethanol yield was 7.52 g ethanol per 100 g DWG (dry basis) and the fermentation was completed within 30 hr. For the bacterial fermentation, DWG was first treated with dilute acid and the syrup, containing the hydrolyzed p entose and starch components, separated from the residual solids. Fermentation of this hydrolysate was completed within 30 hr and the final ethanol concentration was 2.12% w/v. The ethanol yield for the bacterial fermentation was 0.49 g ethanol per g sugar(s) initially present in the hydrolysate, which is 96% of the theoretical yield. DWG is normally used as an animal feed. Pretreating DWG significantly increased the percent of crude protein that was soluble from 4 to 29%, which would have significant impact on the animal nutritional properties of the modified DWG.

Introduction

The U.S. produced 1.77 billion gallons of ethanol in 2001. Ethanol is used as a fuel oxygenate, as required by the Clean Air Act 1990 amendment. Methyl tert-butyl ether (MTBE), which is made from methanol, is the alternative oxygenate. However, MTBE has been identified as a major ground water pollutant (1) a nd t he f ederal g overnment is moving to ban its use in gasoline. The National Energy Bill (currently in conference committee) would waive the requirement for an oxygenate and establish a renewable energy standard, which specifies production of 5 billion gals of ethanol per yr by 2012 (2).

Over 95% of U.S. ethanol comes from dent corn, which is processed by either wet milling or dry grinding (also referred to as dry milling). The use of dry grinding has grown faster relative to wet milling because the former has lower capital cost than the latter. Most plants are being built by farmer cooperatives, often in response to state incentives.

Briefly, the dry grind ethanol process works as follows (for reviews: 3 and 4). Corn is ground in a hammer mill and mixed with water, some a mmonia, recycled liquid recovered from the bottoms of the d istillation, a nd a bout o ne-third of the alpha-amylase. The corn mash is heated to 105-120°C by direct steam injection using a jet cooker. The mash is allowed to cool to 80-90°C, the rest of the alpha amylase is added, and held at that temperature for 30 min. Next, the mash is cooled to 30°C and the pH adjusted to 4.5-5.0 by addition of a mineral acid. The liquefied corn mash is simulatenously saccharified with glucoamylase and fermented (SSF) with *S. cerevisiae*. The fermentation lasts 40-48 hr and the final expected ethanol concentration of the beer is 12-14% v/v. Ethanol is removed by distillation.

In addition to ethanol, dry grind plants produce the following co-products: carbon dioxide, condensed or dried distillers' soubles, distillers wet or dry grain, and distillers dried grains with soubles (*5* and *6*). Carbon dioxide is produced during the fermentation, but because of its low selling price it is collected and sold by only a few of the dry grind ethanol processors. The other products are produced from the whole stillage that remains after the ethanol has been distilled from the fermentation broth. The whole stillage is centrifuged or screened to yield distillers wet grain (more dense material) and thin stillage. The distillers wet grain (DWG) is sometimes sold as is, but only to local farmers because of its short shelf-life. More often, it is combined with condensed thin stillage, dried, and sold as distillers dried grains with soubles (DDGS). When the thin stillage is dried separately, it is sold as distillers dried soubles. (*5*).

The profitability of dry grind facilities is dependent on the selling price of both ethanol and DDGS. However, the price commanded for DDGS is already falling with expanded ethanol and DDGS production; 16 lb of DDGS is produced per bushel of fermented corn (*3*). Tripling ethanol production will, at the least, triple DDGS production. For this reason, it would be desirable to find alternate uses and/or markets for DDGS.

One alternate use would be to convert DDGS into fuel ethanol. DDGS contains approximately 40% carbohydrates, which after being converted to free sugars, would be available for fermentation to ethanol. Furthermore, removing this fiber from DDGS would lower the fiber content and raise the relative protein content. Animal feeds with higher protein contents command a premium price compared to those with lower protein contents. Lowering the fiber content, might also allow for the modified DDGS to be marketed to non-traditional markets. DDGS is currently marketed as a animal feed for beef and dairy herds (*5*). It has been estimated that an additional 133 million gal per yr could be produced by converting DDGS to ethanol (*3*).

However, converting DDGS to ethanol will require modifying the current dry grind plant. A modified dry grind process that includes conversion of DDGS fiber into ethanol is shown in Fig 1. The modified process consists of two additional steps: pretreatment and a second SSF. Pretreatment prepares the cellulose for enzymatic saccharification and hydrolyzes the other carbohydrates (i.e. residual starch and hemicellulose) into free sugars. The cellulose is saccharified enzymatically by cellulase. Fermentation of the fiber hydrolysate is more complicated than with the corn mash. The fiber contains a variety of carbohydrates including: residual starch, hemicellulose, and cellulose (*7*). Hydrolyzing corn hemicellulose produces a mixture of sugars including arabinose, galactose, and xylose (ibid). *S. cerevisiae* does not ferment either arabinose or xylose. For this study, these sugars along with glucose were converted to ethanol using a recombinant ethanologenic *E. coli* strain FBR5,

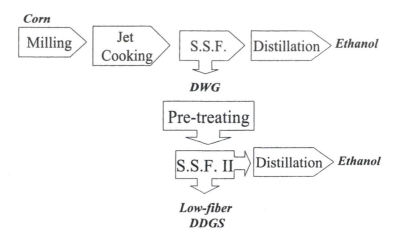

Figure 1. Schematic of corn dry grind process with fiber conversion to ethanol

which was developed by our laboratory *(8* and *9)*. This strain has been metabolically engineered to convert a wide spectrum of sugars to ethanol. Typical ethanol yields are 94% or greater of theoretical (ibid). However, some mills would not want to use recombinant organisms and, so, fermentations were also carried out using *S. cerevisiae*. For the yeast fermentations, only the cellulose and residual starch were converted into ethanol.

Materials and methods

Bacterial strains, plasmid, growth media, and reagents

Samples of DWG were received from Broin Companies (Sioux Falls, SD), Archer Daniels Midland Company (Decatur, IL), and James B. Beam Distilling Co. (Clermont, KY). A sample of corn germ meal was received from Archer Daniels Midland Company (Decatur, IL). Samples were received undried and stored at –20°C. Enzymes were supplied by Novozyme (U.S. office: Franklinton, NC) and included: cellulase (Celluclast® 1.5L; 48 International Filter Paper Units per ml), beta-glucosidase (Novozym®188; 66.8x10^3 I.U./ml), and glucoamylase (AMG300L). Sugars were purchased from Sigma (St. Louis, MO) and all other chemical and media regents from Fisher Scientific (Fairview, NJ).

Media and protocols for routine maintenance of *E. coli* strain FBR5 have been previously described *(8)*. *S. cerevisiae* (Y-2034, NCAUR Culture Collection, Peoria, IL) was stored in 50% v/v glycerol stocks at –80°C. The yeast culture was routinely maintained on YPD (10 g/l yeast extract, 20 g/l peptone, and 20 g/l dextrose with 20 g/l Difco Agar added for solid medium) and incubated at 32°C.

Compositional Analysis of Biomass

Each sample was analyzed for moisture, carbohydrate, oil, and protein contents. Moisture was measured by drying the samples at 105°C until samples had reached a stable weight. Oil was measured using AOAC method 920.39 and protein by AOAC method 976.06, which is based on measuring total nitrogen. Starch was determined as previously reported (*10*). Arabinose and xylose were determined using trifluoacetic acid treatment as described previously (*11*). Cellulose was determined using ASTM method E1758-95. Samples were analyzed for oil, protein, and starch by Analabs (Fulton, IL). Feed analysis of DWG was performed by Dairy One (Cornell, NY).

Optimizing Acid Loading

DWG was pretreated by dilute-acid hydrolysis. The amount of acid added per g biomass was optimized for complete hydrolysis of the hemicellulose and subsequent enzymatic hydrolysis of the cellulose. Eight grams biomass (wb, wet basis) was diluted with a combined volume of 12 ml water and sulfuric acid (0 – 12 g sulfuric acid per 100 g biomass (dry basis, db) to give a solids loading of 14.4 % (wt of biomass, db/ total wt). The mixture was placed in stainless steel pipe reactors (40 ml working volume), which were placed in a fluidized sand bath (Model 01187-00 bath and 01190-72 temperature controller, Cole-Parmer, Vernon Hills, IL). The mixture was heated to and kept at 150°C for 10 min before being quickly cooled in a water bath. The internal reactor temperature was monitored using a thermocouple probe inserted into one of the pipe reactors. The pretreated material was transferred to a test tube, neutralized with Ca(OH)$_2$ to a pH of 4.5 and citric acid buffer (pH = 4.8, 50 mM) added along with cellulase (1.0 %v/v), beta-glucosidase (1.0% v/v). Thymol (0.05% v/v) was added to prevent microbial contamination. The biomass samples were incubated at 45°C with agitation for 48 hr in a water bath (Dubnoff Metabolic Shaking Incubator, Precision Scientific Inc., Chicago, IL). The hydrolysis reactions were sampled at 24 and 48 hr for sugar concentrations.

Simultaneous Saccharification and Yeast Fermentation

The biomass was pretreated in a similar manner as described above for the cellulase hydrolysis experiments with a 3.2% w/w sulfuric acid loading. For SSF, 32 g wet or 11.8 g dry basis (db) of pretreated material was placed in a 125 ml Erlenmeyer Flask to which the following was added: cellulase (0.26 %v/v), beta-glucosidase (0.26 %v/v), glucoamylase (0.036% v/v), and 5%v/v of a 20x Y.P. stock. The SSF was started by inoculating to an initial O.D.600nm of 0.5 from a *S. cerevisiae* seed culture. The beginning solids for the SSF, including all additions, was 11% w/w. The flasks were capped with a rubber stopper, pierced with a 22g needle to exhaust CO_2. The cultures were incubated at 32°C and agitated at 150 rpm (Refrigerated Innova® Shaker, New Brunswick, Edison, New Jersey) for 70 hr. The fermentations were sampled each day for glucose and ethanol concentrations.

The yeast inoculum was started from a glycerol stock (-80°C). The stock was transferred to solid YPD and incubated for 24 hr at 30°C. A single colony was transferred to a 50 ml Erlenmeyer flask containing 10 ml of YPD and subsequently transferred to a 250 ml Erlenmeyer flask containing 50 ml YPD supplemented with an extra 30 g/l glucose. Both liquid cultures were incubated at 32°C and agitated at 200 rpm.

Bacterial Fermentations

For bacterial fermentations, the hydrolysates were prepared as follows. The biomass, either DWG or corn germ meal, was dried at 65°C for approximately 24 hr in a convection oven and ground with a coffee mill. The corn fiber was mixed with 1% v/v H_2SO_4 solution at a ratio of 1.2 g (db) biomass to 5.0 ml, placed in a shallow Pyrex® dish, covered with aluminum foil, and heated at 121°C for an hour. After being allowed to cool, the liquid was separated from the solids using cheesecloth. Next, the recovered liquid portion was treated as follows: (1) adjusted to a pH of 10 by adding $Ca(OH)_2$, (2) 1 g/l of sodium sulfite was added, (3) warmed to 90°C and incubated at this temperature for 30 min, and (4) neutralized with H_2SO_4 to a pH of 7.0. Following neutralization, the resulting precipitates, including gypsum, were removed by centrifugation (10,000 rcf, 15 min). The recovered liquid was filter sterilized through a 0.22 μm^2 membrane filter.

Bacterial fermentations were carried out in mini-bioreactors with automatic pH control that were constructed and operated essentially as described previously (*8* and *12*). Each 500-ml Fleaker® culture vessel contained 170 ml of hydrolysate supplemented with 20 ml of a 10x LB solution (10 g/l tryptone and 5 g/l yeast extract) and antifoam 289 (0.1 ml/l). Nitrogen was bubbled through the medium for 30 min subsequent to inoculation to remove oxygen. The fermentation vessels were each inoculated with a 5% v/v inoculum from an anaerobic culture grown over-night at 37°C. Fermentations were run at 35°C and stirred magnetically with 1 x 1 inch "X" shaped stir bars at 300 rpm. The pH was set at 6.5 and maintained by addition of a concentrated base solution (4 N KOH). Ethanol, sugars, organic acids, and OD_{550} were determined periodically with 1.5 ml samples of cultures. Each experiment was run in duplicate.

Analytical procedures

Activities for cellulase (FPU/ml) and beta-glucosidase (IU/ml) were measured by the methods described in *13* and *14,* respectively. Optical densities (1 cm light path) of cultures were monitored on a Beckman DU-640 Spectrophotometer (Fullerton, CA) at 550 nm. Concentrations of sugars and ethanol were determined by high-pressure liquid chromatography (HPLC) using an Aminex HPX-87H column (Bio-Rad, Richmond, CA) and refractive index detector. Samples were run at 65°C and eluted at 0.6 ml/min with 5 mM sulfuric acid.

Calculations

Ethanol yields and productivities for the fermentations were determined as previously described (*15*). Ethanol yields for the DWG fermentations have also been reported on a per bushel of corn processed basis. The ethanol yield equation, which is similar to those derived in *16*, is shown below.

Ethanol(gal/bushel) = Dry Biomass Yield(lb/bushel corn) x Carbohydrate
  Yield(lb/lb biomass) x 1.11 (lb sugar/ lb carbohydrate) x Fermentation
  Yield(lb ethanol/lb fermented sugar) ÷ 6.58 (lb ethanol/ gal ethanol)

The dry biomass yield for DWG (db) was assume to be 8 lb/bushel. Data on carbohydrate yield of anhydrous sugar(s) from DWG (db) is in Table I. The fermentation yield for a theoretical ethanol yield is 0.51 lb/lb and actual ethanol yield are listed in Table III.

Results and Discussion

Composition of distillers wet grains

DWG samples received from Broin Companies (Sioux Falls, SD), Archer Daniels Midland Company (Decatur, IL), and James B. Beam Distilling Co. (Clermont, KY) were analyzed for carbohydrates, oil, and protein (Table I). The samples represent the spectrum of processes used to produce DWG: Broin is a dry grind fuel ethanol producer, Jim Beam is a whisky maker, and the ADM sample came from a modified wet mill, which includes a germ removal step. The compositions were all similar: total carbohydrates were 34.9-40.5% w/w db, oil 7.8-10.9% w/w, and protein 30-38.3%w/w (Table I). Variations in protein contents among the DWG samples could either reflect differences in

Table I. Composition of Distillers Dried Grains (%w/w)

Source Carbohydrate	Starch Glucose[1]	Cellulose Glucose[1]	Hemicellulose Arabinose[1]	Xylose[1]	Oil	Protein
Jim Beam	1.2 ± 0.0^{2}	13.5 ± 0.4	12.1 ± 0.4	17.9 ± 0.5	9.9 ± 0.4	30.0 ± 0.0
ADM	4.2 ± 0.1	13.9 ± 1.6	10.1 ± 1.0	14.8 ± 1.5	7.8 ± 0.1	38.3 ± 0.4
Broin	2.6 ± 0.1	13.8 ± 0.5	6.9 ± 1.2	12.8 ± 1.9	10.9 ± 0.1	36.1 ± 0.4

[1]anhydrous form of sugar

[2]each determination was done in duplicate

yeast growth or amount of nitrogen source added to the fermentation. There was little residual starch, which is expected because most of the starch would remain with the liquid fraction, which was not used in this study. The components measured a ccount f or 8 2-89% o f t he d ried m aterial, t he r esidual material (not tested for) includes ash, extractables, lignin, and lipids.

Pretreatment and SSF of DWG using *S. cerevisiae*

The carbohydrates present in DWG were converted to free sugars using a combined dilute acid pretreatment and cellulase and fermented to ethanol using *S. c erevisiae*. First, a series of experiments were conducted to determine the optimal amount of acid to add for pretreatment. A successful pretreatment will completely hydrolyze the hemicellulose component of the fiber and allows for efficient enzymatic conversion of cellulose to glucose. The DWG was pretreated with dilute sulfuric acid at 150°C as suggested in (7) for corn fiber. Their protocol was optimized for DWG by varying the acid loading from 0–12.8% g sulfuric acid per g biomass (db). The glucose yield reached its maximum value at an acid loading of 1.6% w/v (Fig 2). Maximum yields for arabinose and xylose occurred at 1.6% w/v and 6.4% w/v (Fig. 2). Interestingly, heating, without acid, was sufficient to recover 80% of the glucose.

For the yeast fermentations, DWG was treated at a high loading solids (14.4% wt biomass (db) / total wt) and an acid loading of 3.2% wt/wt acid loading. The highest possible solid loading was used, such that the material formed a mixable slurry (data not shown). The recovered material was neutralized, mixed with cellulase, beta-glucosidase, and glucoamylase (to ensure complete recovery of the starch fraction) and inoculated with *S. cerevisiae*. The cellulase loading was 15 FPU/g cellulose. The native beta-glucosidase activity of the cellulase mixture was supplemented because adding extra activity has been reported to enhance the rate of SSF (17). The combined hydrolysis and fermentation was competed by 30 hr (Fig. 3) and the final ethanol concentration was 0.84±0.04% w/v. The ethanol yield was 92±4% of theoretical based upon the beginning glucose concentration in the fermentation medium. Assuming 8 lb of DWG (db) is recovered per bushel of corn, the fermentation results suggests an a dditional 0 .108 g al o f e thanol can be gained by fermenting DWG with *S. cerevisiae*. The theoretical yield, based upon the glucose composition of DWG, is 0.113 gal per bushel corn.

Pretreatment and fermentation of DWG using *E. coli* FBR5

Fifty-four percent of the carbohydrates present in DWG are in the form of pentoses, which *S. cerevisiae* does not ferment to ethanol. We have developed a recombinant *E. coli* strain that is capable of fermenting arabinose, glucose, and xylose into ethanol (9). The same pretreatment protocol was used for these

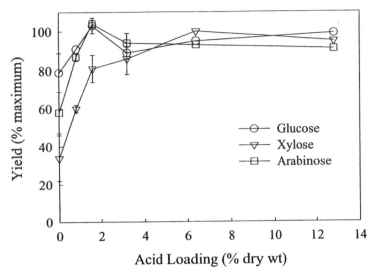

Figure 2. Sulfuric acid loading acid was optimized for pretreating DWG for conversion to monomeric sugars. Each point is average of duplicate runs.

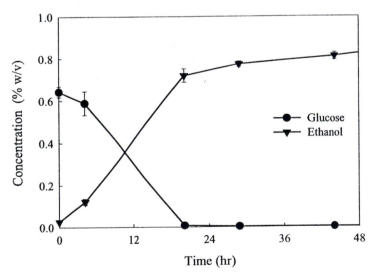

Figure 3. SSF of pretreated DWG with *S. cerevisiae*. Fermentations were performed in duplicate.

experiments as had been used previously for corn fiber, received from a corn wet milling plant (ibid). Unlike the protocol described above for the *S. cerevisiae* fermentations, only the liquid portion of the pretreated material was fermented because the bioreactor systems needed to carry out the fermentations were unable to effectively mix high solids. Likewise no cellulase was added because cellulose partitions with the solids. *E. coli* FBR5 readily fermented the hydrolysates prepared from Broin DWG (Fig. 4). The fermentations were completed in approximately 36 hr and the yields were 96% of theoretical (data not shown). Control cultures, using reagent grade sugars, fermented slightly faster than the hydrolysates, suggesting the presence of some microbial inhibitors. Even though the cellulose was not converted to ethanol, the maximum ethanol concentration of the DWG fermentation was 2.5x greater than that of the *S. cerevisiae*. Based upon the fermentation results, an additional 0.056 gal ethanol per bushel of corn could be made using FBR5. The yield could be readily increased if the recovered solids from the pretreatment had been washed, which was not done to preserve the sugar concentration. If the aforementioned ethanol yield were corrected for the liquid left with the cake, the yield would double to 0.11 gal ethanol per bushel, which is 72% of the maximum possible based upon converting all of the carbohydrates in DWG, except cellulose.

Effect of Pretreatment on DWG feed quality

A major concern with converting the DWG to ethanol is the value of the remaining unfermented residue as an animal feed. DWG is enriched for protein because of the large amount of *S. cerevisiae* produced during the fermentation. Pretreatment of the material at low pH and high temperature could encourage Maillard type reactions, which could reduce the value of the feed. To address this concern, DWG was pretreated at 150°C for 10 minutes, with and without added acid, and analyzed for its feed value (Table II). Acid detergent insoluble crude protein (ADICP), used as a measure of unavailable protein content, only increased from 5.2% (untreated) to 6.2% (pretreated). The major effects on protein were increases in digestible and soluble protein for use in the rumen. These two changes are significant and will alter how the pretreated DWG would be marketed relative to traditional DWG . As expected, neutral detergent fiber (NDF) decreased by one-half when the material was pretreated; NDF is related to the amount of intact hemicellulose.

If the fiber could be removed subsequent to the fermentation step, it could be pretreated in the absence of the yeast protein. There are several ongoing projects for incorporating fiber removal in a dry grind process. One promising technology is the Quick Fiber process (*18*). In this process, the corn is given a quick soak (12 hr) in lactic acid (0.5% w/v) before being passed through a germ mill. The fiber and germ are removed together, dried, and the fiber separated from the germ by air aspiration. While quick fiber was not available for fermentation, we have previously described fermentation of corn fiber, an

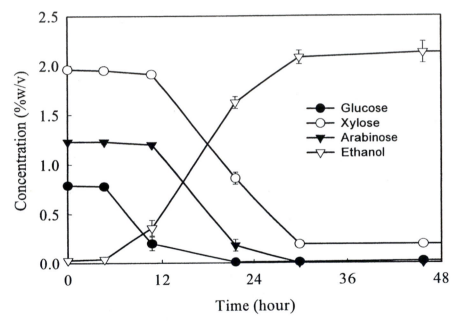

Figure 4. Fermentation of DWG liquid hydrolysate with *E. coli* FBR5. Fermentations were performed in duplicate.

Table II Feed Analysis of DDG pretreated with dilute and/or just heat

Pretreatment	CP[1] % w/w	SP[2] % CP	DP[3] % CP	ADICP[4] %CP	ADF[5] % CP	NDF[6] % w/w
Control	34.8	4	23	5.2	19.7	34.9
Heat	33.8	14	37	7.2	19.3	33
Heat and Acid	34.8±1.5[7]	29±3	59.5±4.5	6.15±0.45	13.6±0.1	18.75±0.25

[1]crude protein, [2]soluble protein, [3]degradable protein, [4]acid detergent insoluble protein, [5]acid detergent fiber, [6]neutral detergent fiber, [7]results for this treatment based upon duplicate trails

Table III: Fermentation results for various fibrous feedstocks using ethanologenic strain FBR5[1]

Feedstock	Initial Sugar Concentrations			Maximum Ethanol	Ethanol Yield	Ethanol Productivity	Reference
	Arabinose % w/v	Glucose % w/v	Xylose % w/v	% w/v	g/g	g/l/h	
DWG[2]	0.79	1.96	1.23	2.12±0.05	0.49±0.01	0.71±0.01	this study
Germ	1.63	1.08	1.44	2.19±0.03	0.50±0.00	0.56±0.01	this study
Corn Fiber	2.00	2.80	3.70	3.74±0.01	0.46±0.00	0.77±0.05	Dien et al, 2000

[1]Each result based upon duplicate fermentations, [2]Broin DWG

analogous product produced by wet-milling (9). For this study, we further studied the fermentation of corn germ, the other source of fiber in the kernel.

Results from these fiber fermentations with strain FBR5 are summarized in Table III. Corn fiber appears to be the most promising substrate for fermenting to ethanol based upon final ethanol concentration (Table III). Corn fiber hydrolysates gave nearly double the final ethanol concentration of hydrolysates prepared from either DWG. Corn fiber also has the highest carbohydrate content (70%w/w) among these feedstocks (Table I and 16). The corn fiber was also easier to pre-treat than the DWG because it absorbed less water (data not shown). The corn germ meal gave a final ethanol concentration similar to that of the DWG.

Fermenting DWG with either *S. cerevisiae* or *E.coli* FBR5 has the capability of increasing the ethanol yield form a bushel of corn. However, pretreating DWG does influence the protein quality and property of this biomass, which will be significant when selling the recovered material for animal feed. The protein quality could be preserved if the dry grind process were modified for fiber recovery subsequent to the starch fermentation. Furthermore, results from fermenting germ meal and previously reported results for corn fiber demonstrate that *E.coli* FBR5 readily ferments hydrolysates prepared from either of these sources of corn associated fiber.

References

1. Anonymous. 1999. E.P.A Blue Panel Report., EPA420-R-99-021.
2. Brown, R. 2002. *Chem Market Reporter*, May 6, 2002.
3. Dien, B.S., R.J. Bothast, and N.N. Nichols. 2002a. *Int Sugar Jnl.* 104:1241, pp 204-211.
4. Elander, R.T. and V.L. Putsche. 1996. In: Handbook on bioethanol. Ed. C.E. Wyman. Publisher: Taylor and Francis, Washington D.C. 329-350.
5. Akayezu, J.M., J.G. Linn, R.H. Summer, and J.M. Cassady (1998) *Feedstuffs.* 70:52
6. Murtagh J.E. 1999. *In.* Jacques, K., Lyons, T.P., D.R. Kelsall. 1999. The Alcohol Textbook; A reference for the beverage, fuel and industrial alcohol industries. 3rd edition. Nottingham University Press. Nottingham, U.K.
7. Grohmann, K. and R.J. Bothast. 1997. *Process Biochem.* 32:405-415.
8. Dien B.S., R.B. Hespell, H Wyckoff and R.J. Bothast. 1998. *Enzyme Microbial Technol* 23: 366-371.
9. Dien, B.S., N.N. Nichols, P.J. O'Bryan, and R.J. Bothast. 2000. *Appl Biochem Biotechnol.* 84-86:181-196.

10. Dien, B.S., R.J. Bothast, L.B. Iten, L. Barrios, and S.R. Eckhoff. 2002b. *Cereal Chem.* (in press).
11. Dien, B.S., R.B. Hespell, L .O. Ingram a nd R .J. Bothast. 1997. *World J Microbiol Biotech* 13: 619-625.
12. Beall D.S., K. Ohta and L.O. Ingram. 1991. *Biotechnol Bioeng* 38: 296-303.
13. Ghose, T.K. 1987. *Pure & Appl. Chem.* 59:257-268.
14. Hespell, R.B., R. Wolf, and R.J. Bothast. 1987. *Appl Environ Microbiol.* 53:12, p. 2849-2853.
15. Nichols, N.N., B.S. Dien, R.J. Bothast 2001. *Appl Microbiol Biotechnol*, 56 (1-2): 120-125.
16. Gulati M, K Kohlmann, MR Ladisch, R Hespell and RJ Bothast. 1996. *Bioresource Technol.* 58: 253-264.
17. Grohmann, K. 1993. In: Bioconversion of forest and agricultural plant residues. Ed. J.N. Saddler. Publisher: C.A.B. International, Wallingford, U.K. pp 183-210.
18. Singh V. and S.R. Eckhoff. 1996. *Cereal Chem.* 73(6):716-720.

Agricultural Processes
in Green Chemistry

Chapter 7

Chloroplast Bioengineering 1: Photosynthetic Efficiency, Modulation of the Photosynthetic Unit Size, and the Agriculture of the Future

Constantin A. Rebeiz, Vladimir L. Kolossov, and Karen K. Kopetz

Laboratory of Plant Biochemistry and Photobiology, NRES, University of Illinois, Urbana, IL 61801

The world population of about 6 billion is expected to increase to 9 billion by the year 2030. It may reach 18 billion by the end of the century. Worldwide, there has been a progressive decline in cereal yield, and at present, the annual rate of yield increase is below the rate of population increase. Since it will be difficult to increase the land area under cultivation without serious environmental consequences, the increased demand for food and fiber will have to be met by higher agricultural plant productivity. Plant productivity depends in turn on photosynthetic efficiency. We have reason to believe that agricultural productivity can be significantly increased by alteration of the photosynthetic unit size. On the basis of recent advances in the understanding of the chemistry and biochemistry of the greening process and significant advances in molecular biology, we believe that alteration of the PSU size has become a realistic possibility.

Life in the biosphere is carbon based. All molecules needed for life are made up of a carbon skeleton which is complemented by organic elements such as O, H, N, and inorganic elements such as K, P, Ca, Fe, etc. Carbon, O and H of organic compounds originate in CO_2 and H_2O. The carbon skeleton is assembled via the process of photosynthesis which essentially converts solar energy into chemical energy. Nitrogen originates in NH_3 and inorganic elements originate in the rocks of the biosphere and are incorporated into the carbon skeleton via enzymatic reactions. Chemical energy consists of the covalent bond energy embedded into the carbon-carbon skeleton as well as the high energy bonds of ATP and NADPH which are formed during the process of photosynthesis.

The carbon cycle essentially describes how photosynthesis supports organic life in the biosphere. The carbon skeleton formed via the process of photosynthesis is converted into the simple and complex food consumed by organic life. The needed energy for enzymatic interconversions and biosynthetic processes is provided by ATP and NADPH. The organic matter of dead biota is converted in turn into CO_2, H_2O, and inorganic elements by bacterial activity, then the cycle repeats itself all over again.

At issue then, is whether agricultural productivity at today's levels of photosynthetic efficiency is efficient enough to feed a growing world population. Indeed the world population of about 6 billion is expected to increase to 9 billion by the year 2030 and may easily reach 18 billion by the end of the twenty first century.

Agricultural Productivity and Photosynthetic Efficiency

Since plants form food by conversion of solar energy, CO_2, and H_2O into chemical energy via the process of photosynthesis, it ensues that agricultural productivity depends in turn upon photosynthetic efficiency. Let us therefore briefly dissect the components of photosynthetic efficiency.

Photosynthetic efficiency is controlled by intrinsic and extrinsic factors (1). Extrinsic factors include the availability of water, CO_2, inorganic nutrients, ambient temperature, and the metabolic and developmental state of the plant. The most important intrinsic factor is the efficiency of the photosynthetic electron transport system (PETS). The latter is driven by two photochemical reactions that take place in membrane-bound photosystem I (PSI), and PSII chlorophyll (Chl)-protein complexes.

The Primary Photochemical Act of PSI and PSII

Conversion of solar energy into chemical energy is the results of two photochemical acts that take place in PSI and PSII. The primary photochemical act of PSII is initiated by the absorption of light by antenna Chl *a* and *b*. The absorbed photons are conveyed to special Chls in the PSII reaction center. There, the light energy is used to generate a strong oxidant Z^+ which liberates oxygen from water. It also generates a weak reductant Q^- which together with plastoquinone electron acceptor pools serve for temporary storage of the electrons extracted from water.

The primary photoact of PSI is also initiated by the absorption of light by antenna Chl *a* and *b*, and here too, the absorbed photons are conveyed to special Chls in the PSI reaction center. There the light energy generates a weak oxidant $P700^*$ which receives electrons from the plastoquinone pools via cytochrome f and plastocyanin. It also generates a strong reductant A_0 which donates electrons to $NADP^+$ via a series of electron carriers and converts it to NADPH. The photochemical acts of PSII and PSI, and the flow of electrons between PSII and PSI are depicted in Fig. 1.

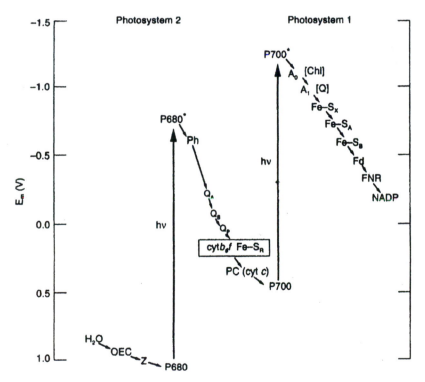

Fig. 1. The Z scheme for electron transport in oxygenic photosynthesis (Reproduced from reference 2. Copyright 2002 Blackwell Publishing.)

Conversion of CO_2 into Carbohydrates

During electron and proton flow, energy rich ATP and NADPH are formed. The energy of NAPDH and ATP is used for the enzymic conversion of CO_2 into carbohydrates which are in turn converted into a variety of organic molecules. In summary the efficiency of food formation by green plants depends to a great extent on the efficiency of NADPH and ATP formation which depends in turn on the efficiency of the PETS.

The rest of this chapter will therefore be devoted to a discussion of the efficiency of the PETS and possible alterations in the circuitry of the chloroplast that may lead to a higher efficiency of the PETS and higher plant productivity under field conditions.

Theoretical Maximal Energy Conversion Efficiency of the PETS of Green Plants

This discussion is essentially extracted from a 1975 RANN report (*1*). At the maximal quantum efficiency of one, two photons are required to move one electron across the potential difference of about 1.25 V between Z^+ and A_0. The maximal efficiency of the photochemical reactions leading to the formation of Z^+ and A_0 is given by

$$E = 1.25 \text{ eV}/ 2 \text{ h}\nu \qquad (1)$$

Where,

E = efficiency of PETS
eV = Energy units in electron volts
hν = Energy of the absorbed photon in eV

Since the red 680 nm photons absorbed by PSI and PSII have an energy of 1.83 eV, it ensues from Eq 1 that

$$E = 1.25 \text{ eV}/2*1.83 \text{ eV} = 0.34 \text{ eV} \qquad (2)$$

Therefore the absolute maximal efficiency of the PETS under red light is

$$(0.34 \text{ eV}/1.25 \text{ eV})*100 = 27\% \qquad (3)$$

Under natural white light, although the Chl concentration in photosynthetic membranes is high enough to result in near total absorption of all incident photosynthetically active photons between 400 and 700 nm, under normal weather conditions, these photons represent only about 44.5 % of the total incident solar radiation. Therefore under these circumstances, the possible overall maximal energy conversion efficiency amounts to:

$$(27\%*44.5\%)/100 = 12\% \tag{4}$$

Actual Energy Conversion Efficiency of the PETS of Green Plants under Field Conditions

Under field conditions however, the average net photosynthetic efficiency results in a net agricultural productivity in the range of 2-8 tons of dry organic matter per acre per year (*1*). This corresponds to a solar conversion efficiency of 0.1 to 0.4% of the total average incident radiation. Therefore the discrepancy between the 12% maximal theoretical efficiency of the PETS, and the agricultural photosynthetic efficiency observed under field conditions ranges from

$$(12\%/0.4\%)*100 = 3000\%, \text{ to} \tag{5}$$

$$(12\%/0.1\%)*100 = 12000\% \tag{6}$$

Molecular Basis of the Discrepancy Between the Theoretical Maximal Efficiency of the PETS and the Actual Solar Conversion Efficiency of Photosynthesis Under Field Conditions

The discrepancy between the 12% theoretical maximal efficiency of the PETS and the actual 0.1-0.4% solar conversion efficiency of photosynthesis observed under field conditions can be attributed to (a) factors extrinsic to the PETS, and (b) to intrinsic rate limitations of the PETS (*1*).

Contribution of Extrinsic PETS Parameters to the Discrepancy Between the Theoretical Photosynthetic Efficiency of 12% and the Actual Photosynthetic Field Efficiency of 0.1-0.4%

Photosynthetic efficiency under field conditions is directly or indirectly affected by extrinsic factors such as ambient weather conditions, availability of water, CO_2, and inorganic nutrients, as well as the metabolic and developmental state of the plant. Some of those factors are under human control while others are not. They do contribute nevertheless, to the variation in photosynthetic efficiency under field conditions. The rest of this discussion will focus upon the impact of intrinsic factors that affect the PETS and photosynthetic efficiency.

Contribution of Intrinsic PETS Parameters to the Discrepancy Between the Theoretical Photosynthetic Efficiency of 12% and the Actual Photosynthetic Field Efficiency of 0.1-0.4%

The 12% theoretical efficiency of the PETS assumes that under natural conditions, PSI and PSII operate at a maximal quantum efficiency of ONE. In other words, it is assumed that every absorbed photon is completely converted into energy without losses(*1*).

Using the conventional figure of 200 light harvesting Chl molecules per reaction center (RC) per PS, *i. e.* for a photosynthetic unit (PSU) size of 200 per PS, under the moderate light int ensities of a shady sky (about 1/10 of full sunlight), each RC would receive about 200 photons per second (s) (*1*). In other words, each RC would receive about 200 hits or excitons per s. Under these conditions, in order to maintain a quantum efficiency of ONE, the slowest dark reaction of the entire PS must have a turnover rate of 200 per s(*1*).

Under full sunlight, which is about 10 fold higher than in the shade, the turnover rate of the limiting dark-reaction should be 200*10 = 2000 per s. This turnover rate corresponds to a rate of O_2 evolution of about 9000 µmoles of O_2 per mg Chl per hour (h). Yet, the maximal rate of O_2 evolution observed during a Hill reaction, which results in the oxidation of H_2O and the release of O_2, under saturating light intensities, and other optimal conditions, rarely exceeds 5-10% of the above value. In other words, it is equal to the optimal rate of O_2 evolution observed in the shade (*1*).

Furthermore extensive kinetic studies have demonstrated that the rate limiting steps of the PETS do not reside in the initial photochemical reactions that take place in the RC, but reside within the redox-carriers, *i. e.* the electron transport chains connecting PSII to PSI. The discrepancy between the capacity of the photon gathering apparatus, *i. e.* the antenna Chl-protein complexes and the capacity of the rate-limiting dark reactions has been named the antenna/PS Chl mismatch (*1*).

Correction of the Antenna/PS Chl Mismatch

The first and most important effect of the antenna/PS Chl mismatch is one of reduced quantum conversion efficiencies at light intensities above shade levels. The second effect relates to the photodestructive effects of the excess photons collected by antenna Chl but not used in the initial photochemical acts. The energy of these unused photons leads to serious photodestruction of the PETS which must be repaired at a cost (*1*).

Early on, the possible correction of the antenna/PS mismatch attracted the interest and curiosity of the photosynthesis community. It was suggested that one way of correcting the mismatch was by reducing the size of the PSU, which may be achieved by growing plants with chloroplasts having less antenna and more RC Chl per unit thylakoid area (*1*). Research performed in the early 1970s failed however in its effort to alter significantly the PSU size in algal cell cultures(*1*).

Now, on the basis of deeper understanding of the chemistry and biochemistry of the greening process, which was achieved during the past 30 years, we have reason to believe that alteration of the PSU has become a realistic possibility.

What Kind of Scientific Knowledge is Needed to Bioengineer a Reduction in PSU size

Thorough and integrated anabolic and catabolic knowledge in the following fields of research is needed for successful research aimed at the bioengineering of a reduced PSU size namely: (a) Chl, (b) lipid, (c) carotenoid, (d) plastoquinone, (e) chloroplast apoprotein, and (f) assembly of pigment-protein complexes. Because of space limitations, the remainder of this discussion will focus on the Chl, and apoprotein components of chloroplasts as well as on the assembly of Chl-protein complexes.

State of the Art in our Understanding of Chl biosynthesis

During the past 30 years, it has become apparent that contrary to previous beliefs, the Chl biosynthetic pathway, is not a simple single-branched pathway, but a complex multibranched pathway that consist of about 15 carboxylic and

88

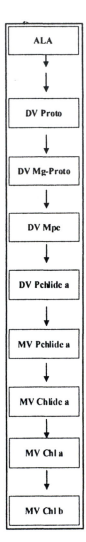

Fig. 2. The single branched Chl biosynthetic pathway. [reproduced from reference (3). Copyright 1994, John Wiley and sons]

two fully esterified biosynthetic routes (2). The single and multibranched carboxylic pathways are briefly discussed below.

The Single-Branched Chl Biosynthetic Pathway does not Account for the Formation of all the Chl in Green Plants

The single-branched Chl biosynthetic pathway is depicted in Fig. 2. It consists of a linear sequence of biochemical reactions which convert divinyl (DV) protoporphyrin IX (Proto) to monovinyl (MV) Chl *a* via DV Mg-Proto, DV Mg-Proto monomethyl ester (Mpe), DV protochlorophyllide *a* (Pchlide *a*), MV Pchlide *a*, and MV Chlorophyllide *a* (Chlide *a*). The salient features of this pathway are (a) the assumption that DV Pchlide *a* does not accumulate in higher plants, but is a transient metabolite which is rapidly converted to MV Chl *a* via MV Pchlide *a*, and (b) that the formation and accumulation of MV tetrapyrroles between Proto and Mpe and DV tetrapyrroles between Pchlide *a* and Chl *a* does not take place (3). All in all, experimental evidence gathered over the past 30 years indicates that only a small fraction of the total Chl of green plants is formed via this pathway (2;4)

The Chl of Green Plants is Formed via a Multibranched Biosynthetic Pathway

Since the 1963 seminal review of Smith and French (5), our understanding of the Chl biosynthetic pathway has changed dramatically. Several factors have contributed to this phenomenon, among which: (a) development of systems capable of Chl and thylakoid membrane biosynthesis *in organello* and *in vitro*, (6-11), (b) powerful analytical techniques that allowed the qualitative and quantitative determination of various intermediates of the pathway (4), (c) recognition that the greening process proceeds differently in etiolated and green tissues, in darkness and in the light, and in plants belonging to different greening groups (12-14), and (d) recognition of the probability that the structural and functional complexity of thylakoid membranes is rooted in a multibranched, heterogeneous Chl biosynthetic pathway (15). Chlorophyll biosynthetic heterogeneity refers in turn either (a) to spatial biosynthetic heterogeneity, (b) to chemical biosynthetic heterogeneity, or (c) to a combination of spatial and chemical biosynthetic heterogeneities (2). Spatial biosynthetic heterogeneity refers to the biosynthesis of an anabolic tetrapyrrole or end product by identical sets of enzymes, at several different locations of the thylakoid membrane. On the other hand, chemical biosynthetic heterogeneity refers to the biosynthesis of an anabolic tetrapyrrole or end product at several different locations of the thylakoid membrane, via different biosynthetic routes, each involving at least one different enzyme. Figure 3 organizes all known carboxylic Chl biosynthetic reactions into a logical scheme made up of 15 different biosynthetic routes.

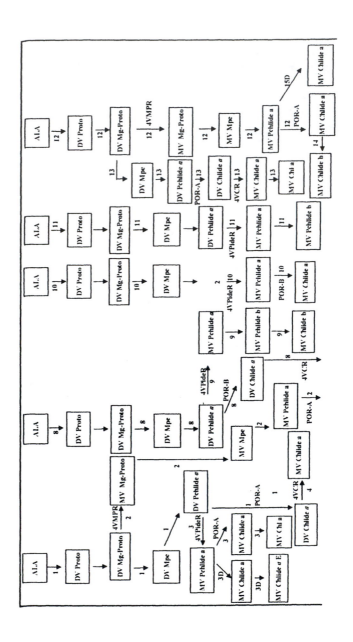

91

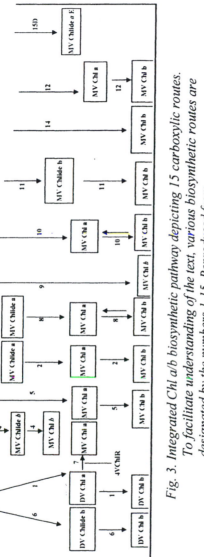

Fig. 3. Integrated Chl a/b biosynthetic pathway depicting 15 carboxylic routes. To facilitate understanding of the text, various biosynthetic routes are designated by the numbers 1-15. Reproduced from http://w3.aces.uiuc.edu/nres/lppbp/greeningprocess/B. biosynthetic Heterogeneity of the Chl Biosynthetic Pathway: an overview.

Thylakoid Apoprotein Biosynthesis

The biosynthesis of thylakoid apoproteins is a very complex phenomenon. Some apoproteins are coded for by nuclear DNA, are translated on cytoplasmic ribosomes and are transported to developing chloroplasts. Other apoproteins are coded for by plastid DNA and are translated on chloroplast ribosomes. A detailed discussion of chloroplast apoprotein biosynthesis is beyond the scope of this discussion. The reader is referred to reference (*16*) for a comprehensive discussion of this topic. For the purpose of this discussion it suffices to say that a PSU is an extremely complex structure that consists of many highly folded thylakoid and soluble proteins as well as membrane-bound pigment protein complexes having different functions in the light steps and dark steps of photosynthesis. A working model of a PSU is depicted in Fig. 4.

Assembly of Chl-Protein Complexes

Success in the bioengineering of smaller PSUs resides in a thorough understanding of how the Chl and thylakoid apoprotein biosynthetic pathways are coordinated to generate a specific functional Chl-protein complex. It is known for example that an apoprotein formed in the cytoplasm or in the chloroplast has to pick up Chl molecules, has to fold properly, and has to wind up in the right place on the thylakoid in order to become a functional Chl-protein complex having a specific role in photosynthesis. What is unknown however is how an apoprotein formed in the cytoplasm or in the chloroplast becomes associated with Chl to become a specific Chl-protein complex of PSI, PSII or light harvesting Chl-protein complex II (LHCII), having a specific function in photosynthesis. We have recently examined three possible models for that scenario, which have been referred to as: (a) the single-branched Chl biosynthetic pathway (SBP)-single location model, (b) the SBP-multilocation model, and (c) the multi-branched Chl biosynthetic pathway (MBP)-sublocation model. The models take into account the dimension of the PSU (*17*), the biochemical heterogeneity of the Chl biosynthetic pathway (*2;4*), and the biosynthetic and structural complexity of thylakoid membranes (*16*). The three models are described below.

Assembly of Chl-Protein Complexes: The SBP-Single Location Model

The SBP-single location model is depicted schematically in Fig. 5. Within the PSU, this model accommodates only one Chl-apoprotein biosynthesis center and no Chl-apoprotein biosynthesis subcenters. Within the Chl-apoprotein biosynthesis center, Chl *a* and *b* are formed via a single-branched Chl

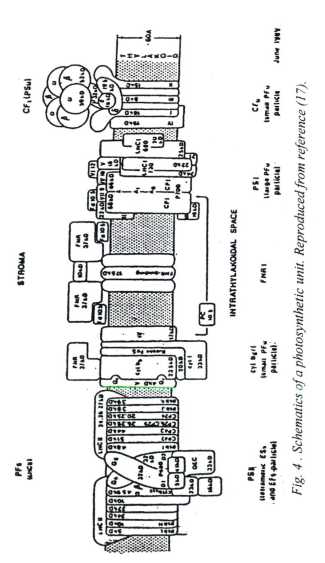

Fig. 4 . Schematics of a photosynthetic unit. Reproduced from reference (17).

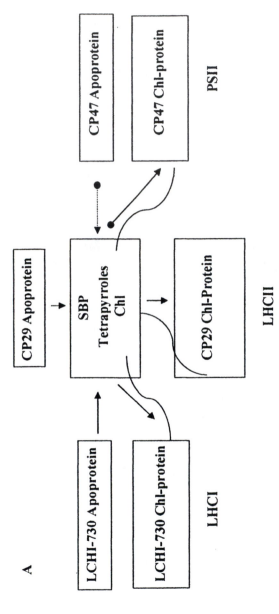

A

LCHI-730 Apoprotein

CP29 Apoprotein

SBP
Tetrapyrroles
Chl

CP47 Apoprotein

LCHI-730 Chl-protein

CP29 Chl-Protein

CP47 Chl-protein

LHCI

LHCII

PSII

Fig. 5. Schematics of the SBP-single location model in a PSU. As an example, the functionality of the model was illustrated with the use of three apoproteins namely CP29, LCHI-730 and CP47. Abbreviations: SBP = single-branched Chl biosynthetic pathway; PSII = photosystem II; LHCII, the major light-harvesting Chl-protein complex of PSII, LHCI, one of the LHC antennae of PSI, CP47 and CP29, two PSII antennae, LHCI-730, the LHC antenna of PSI. Curved lines indicate putative energy transfer between tetrapyrroles and a Chl-protein complex. Adapted from reference (18).

biosynthetic pathway (Fig. 2) at a location accessible to all Chl-binding apoproteins. The latter will have to access that location in the unfolded state, pick up a complement of MV Chl *a* and/or MV Chl *b,* and undergo appropriate folding. Then the folded Chl-apoprotein complex has to move from the central location to a specific PSI, PSII, or Chl *a/b* LHC-protein location within the Chl-apoprotein biosynthesis center over distances of up to 225 Å (4). In this model, it is unlikely to observe resonance energy transfer between metabolic tetrapyrroles and some of the Chl-apoprotein complexes located at distances longer than 100 Å. This is because resonance excitation energy transfer takes place only over distances shorter than 100 Å (*19*).

Assembly of Chl-Protein Complexes: The SBP-Multilocation Location Model

The SBP-Multilocation model is depicted schematically, in Fig. 6. In this model, every biosynthetic location within the PSU is considered to be a Chl-apoprotein thylakoid biosynthesis center. In every Chl-apoprotein biosynthesis location, a complete single-branched Chl *a/b* biosynthetic pathway (Fig. 2) is active. Association of Chl *a* and/or Chl *b* with specific PSI, PSII, or LHC apoproteins at any location is random. In every Chl-apoprotein biosynthesis center, distances separating metabolic tetrapyrroles from the Chl-protein complexes are shorter than in the SBP-single-location model. Because of the shorter distances separating the accumulated tetrapyrroles from Chl-protein complexes, resonance excitation energy transfer between various tetrapyrroles and Chl-apoprotein complexes within each center may be observed. However, formation of MV Mg-Proto (Mp) and its MV methyl ester (Mpe) [*i. e.* Mp(e)] is not observed in any pigment-protein complex. This is because the single-branched Chl biosynthetic pathway does not account for MV Mp(e) biosynthesis.

Assembly of Chl-Protein Complexes: The MBP-Sublocation Model

The SBP-sublocation model is depicted schematically, in Fig. 7. In this model, the unified multibranched Chl *a/b* biosynthetic pathway (*2;4;15*) is visualized as the template of a Chl-protein biosynthesis center where the assembly of PSI, PSII and LHC takes place. The multiple Chl biosynthetic routes are visualized, individually or in groups of one or several adjacent routes, as Chl-apoprotein biosynthesis subcenters earmarked for the coordinated assembly of individual Chl-apoprotein complexes. Apoproteins destined to some of the subcenters may possess specific polypeptide signals for specific Chl biosynthetic enzymes peculiar to that subcenter, such as 4-vinyl reductases, formyl synthetases or Chl *a* and Chl *b* synthetases. Once an apoprotein formed in

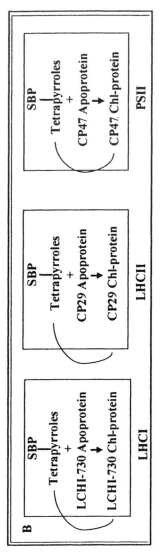

Fig. 6. Schematics of the SBP-single location model in a PSU. All abbreviations and conventions are as in Fig. 5. Adapted from reference (18).

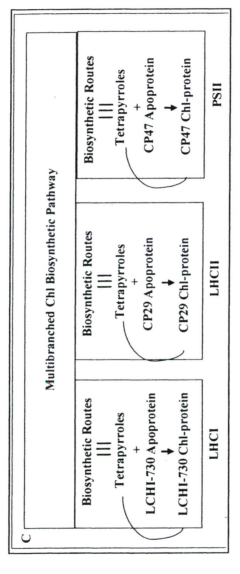

Fig. 7 Schematics of the SBP-single location model in a PSU. All abbreviations and conventions are as in Fig. 5. Adapted from reference (18)

the cytoplasm or in the plastid reaches its subcenter destination and its signal is split off, it binds nascent Chl formed via one or more biosynthetic routes, as well as carotenoids. During pigment binding, the apoprotein folds properly and acts at that location, while folding or after folding, as a template for the assembly of other pigment-proteins. Because of the shorter distances separating the accumulated tetrapyrroles from Chl-protein complexes, within each subcenter, resonance excitation energy transfer between various metabolic tetrapyrroles and Chl is readily observed. In this model, both MV and DV Mp(e) may be present in some pigment-protein complexes, in particular if more than one Chl biosynthetic route is involved in the Chl formation of a particular Chl-protein complex.

Which Chl-Thylakoid Apoprotein Assembly Model is Favored by Experimental Evidence

In order to determine which Chl-thylakoid apoprotein assembly model is likely to be functional during thylakoid membrane formation, we tested the compatibility of the three aforementioned models by resonance excitation energy transfer between anabolic tetrapyrrole intermediates of the Chl biosynthetic pathway and various thylakoid Chl-protein complexes.

Fluorescence resonance excitation energy transfer involves the transfer of excitation energy from an excited donor "D*" to an unexcited acceptor "A" (19-21). The transfer is the result of dipole-dipole interaction between donor and acceptor and does not involve the exchange of a photon. The rate of energy transfer depends upon (a) the extent of overlap of the emission spectrum of the donor and the absorption spectrum of the acceptor, (b) the relative orientation of the donor and acceptor transition dipoles, and (c) the distance between donor and acceptor molecules. The probability of resonance excitation energy transfer between donors and acceptors increases with time but is inversely proportional to the sixth power of the fixed distance separating the centers of the donor and acceptor molecules. It has been estimated that dipole-dipole energy transfer between donor and acceptor molecules may occur up to separation distances of 50 to 100 Å (19).

Resonance excitation energy transfer from three tetrapyrrole donors to the Chl a of various Chl-protein complexes were monitored, namely: from Proto, Mp(e) and MV and DV Pchlide a. DV Proto is a common precursor of heme and Chl. It is the immediate precursor of DV Mg-Proto. As such, it is an early intermediate along the Chl biosynthetic chain. Biosynthetically, it is several steps removed from the Chl end product (2). Mg-Proto is a mixed MV-DV, dicarboxylic tetrapyrrole pool, consisting of DV and MV Mg-Proto (2). It is the precursor of DV and MV Pchlide a. The [(Pchl(ide)] of higher plants consists of about 95% Pchlide a and about 5% Pchlide a ester . The latter is esterified with long chain fatty alcohols at position 7 of the macrocycle. While Pchlide a ester consists mainly of MV Pchlide *a* ester, Pchlide *a* consists of DV and MV Pchlide *a*. The latter are the immediate precursors of DV and MV Chlide *a*. (2). Accumulation of the various tetrapyrrole donors was induced by incubation of green tissues with δ-aminolevulinic acid (ALA) and/or 2,2'-dipyridyl (*22*).

The task of selecting appropriate Chl *a*-protein acceptors was facilitated by the fluorescence properties of green plastids. At 77 °K, emission spectra of isolated chloroplasts exhibit maxima at 683-686 nm (~F685), 693-696 nm (~F695), and 735-740 nm (~F735). It is believed that the fluorescence emitted at ~F685 nm arises from the Chl *a* of LHCII, the major thylakoid LHC antenna, and LHCI-680, one of the LHC antennae of PSI (*17*). That emitted at ~F695 nm is believed to originate mainly from the Chl *a* of CP47 and CP29, two PSII antennae (*17*). That emitted at ~F735 nm is believed to originate primarily from the Chl *a* of LHCI-730, a PSI antenna (*17*). Since these emission maxima are readily observed in the fluorescence emission spectra of green tissues and are associated with definite thylakoid Chl *a*-protein complexes, it was conjectured that they would constitute a meaningful resource for monitoring excitation resonance energy transfer between anabolic tetrapyrroles and representative Chl a-protein complexes. To monitor the possible occurrence of resonance excitation energy transfer between the accumulated anabolic tetrapyrroles and Chl *a*-protein complexes, excitation spectra were recorded at 77°K at the respective emission maxima of the selected Chl *a* acceptors, namely at ~685, ~695, and ~735nm. It was conjectured that if resonance excitation energy transfers were to be observed between the tetrapyrrole donors and the selected Chl *a* acceptors, definite excitation maxima would be observed. These excitation maxima would correspond to absorbance maxima of the various tetrapyrrole donors, and would correspond to the peaks of the resonance excitation energy transfer bands.

The SBP-Single Location Model is not Compatible with Resonance Excitation Energy Transfer Between Anabolic Tetrapyrrole Donors and Chl a-Proteins Acceptors in Chloroplasts

The compatibility of the SBP-single location model with the formation of Chl *a*-thylakoid proteins was tested by monitoring resonance excitation energy transfer between anabolic tetrapyrrole intermediates of the Chl biosynthetic pathway and various thylakoid Chl *a*-protein complexes. Pronounced resonance excitation energy transfer bands from Proto, Mp(e), and Pchl(ide) *a* to Chl *a* ~F685, ~F695, and ~F735 were detected (Table I).

Assignment of *in situ* resonance excitation energy transfer maxima to various metabolic tetrapyrroles was unambiguous except for a few cases at the short wavelength and long wavelength extremes of excitation bands. Contrary to previous beliefs, it was surprising to observe a significant diversity in the various intra-membrane environments of Proto, Mp(e), and Pchl(ide) *a* (18) . This diversity was manifested by a differential donation of resonance excitation energy transfer from multiple Proto, Mp(e) and Pchl(ide) *a* sites to the different Chl *a*-apoprotein complexes. This in turn, is strongly compatible with the biosynthetic heterogeneity of the Chl biosynthetic pathway.

Since resonance excitation energy transfer is insignificant at distances larger than 100 Å (*19*), the detection of pronounced resonance excitation energy transfer from Proto, Mp(e), and Pchl(ide) *a* to Chl *a* ~F685, ~F695, and ~F735 (Table I), indicated that these anabolic tetrapyrroles were within distances of 100

Table I. Mapping of excitation resonance energy transfer maxima to Chl*a* ~F685, Chl *a* ~F695 and Chl *a* ~F735 *in situ*.

Plant Species	Major Donor	Excitation Resonance Energy Maxima to:		
		Chl *a* ~F685	Chl *a* ~F695	Chl *a* ~F735
Cucumber	Proto	410, 415	410	408
Barley	Proto	411	408, 414	407, 411
Cucumber	Mp(e)	422, 432	420, 425	417, 429
Barley	Mp(e)	420, 428	424, 430	426, 432
Cucumber	DV Pchlide *a*	443, 449, 457	436, 442, 453, 463	439, 453, 457, 460
Barley	MV Pchlide *a*	433, 441, 452, 460	438, 444, 449, 463	440, 449, 458

Adapted from reference (*18*).

Å or less of the Chl *a* acceptors. This in turn was incompatible with the functionality of the SBP-single location Chl-thylakoid biogenesis model. Indeed, it can be estimated from published data that the size of the PSU which includes the two photosystems, LHCs, as well as the CF0-CF1 ATP synthase is about 130 x 450 Å (*17*). Most PSU models depict a central cyt b_6 complex flanked on one side by PSI and coupling factor CF1, and on the other side by PSII and LHCII (Fig. 4). With this configuration, the shortest distance between the single-branched Chl biosynthetic pathway and PSI, PSII, and LHCII, in the SBP-single location model would be achieved if the SBP occupied a central location within the PSU. In that case it can be calculated from the PSU model proposed by Bassi (*17*) that the core of PSII including CP29, would be located about 126 Å away from the SBP. On the other hand, LHCI-730 would be located about 159 Å on the other side of the SBP. The centers of the inner and outer halves of LHCII surrounding the PSII core would be located about 156 Å (outer half) and 82 Å (inner half) from the SBP. The detection of pronounced resonance excitation energy transfer from Proto, Mp(e), and Pchl(ide) *a* to Chl *a* ~F685, ~F695, and ~F735 indicates that these anabolic tetrapyrroles are within distances of 100 Å or less of the Chl *a* acceptors. In view of the above considerations it is concluded that the detection of resonance excitation energy transfer between anabolic tetrapyrroles and the Chl *a* of various Chl-protein complexes is not compatible with the functionality of the SBP-single location Chl-thylakoid biogenesis model.

The SBP-Multilocation Model is not Compatible with the Realities of Chl Biosynthesis in Green Plants

Although the SBP-multilocation model is compatible with the detection of resonance excitation energy transfer from Proto and MV Pchlide *a* to Chl *a*-protein acceptors, it is incompatible with (a) our current knowledge of the Chl biosynthetic pathway, and (b) with observed resonance excitation energy transfer from DV Pchlide *a* to Chl *a*-protein acceptors (Table I). Indeed, the SBP does not account for the formation of MV Mp(e) in green plants. Furthermore DV Pchlide *a* is supposed to be a transient metabolite that does not accumulate in higher plants (*23*), yet definite resonance excitation energy transfers from DV Pchlide *a* to Chl *a*-protein acceptors were observed (Table I).

The MBP-Sublocation Model is Compatible with the Realities of Chl Biosynthesis in Green Plants, and with Resonance Excitation Energy Transfer Considerations between Anabolic Tetrapyrrole Donors and Chl a-Protein Acceptors in Chloroplasts

As was mentioned earlier, an overwhelming body of experimental evidence , discussed in (*2*) supports the operation of a multibranched Chl biosynthetic pathway in green plants. Furthermore the MBP-sublocation Chl-protein assembly model requires shorter distances between anabolic tetrapyrrole donors and Chl a-protein acceptors, which leads to the detection of resonance excitation energy transfer between tetrapyrrole donors and acceptors as reported in Table I.

Also, quantum mechanical calculations of resonance excitation energy transfer rates, and distances separating tetrapyrrole donors from Chl a-protein acceptors as well as other considerations discussed in (24) favor the operation of the MBP-sublocation Chl biosynthesis-thylakoid biogenesis model. Table II, reports the calculated distances "R" in Angstroms (Å) separating Proto, Mp(e) and Pchlide a donors from Chl a-Protein acceptors in barley and cucumber chloroplasts at 77°K $in\ Situ$. Distances ranged from a high of 48.3 Å to a low of 11.2 Å (24), which is compatible with the operation of the MBP-sublocation model.

Epilogue

Future research dealing with the bioengineering of smaller PSU sizes will have to use as a working hypothesis the MBP-sublocation Chl a-thylakoid protein biosynthesis model. The first order of business will have to deal with the determination of which Chl biosynthetic routes gives rise to PSI, PSII and LHCII Chl-protein complexes. The greening process may then be manipulated to bioengineer genetically modified plants with a smaller PSU, $i.\ e.$ with more PSU units having less antenna Chl per unit thylakoid area. Nevertheless this type of agriculture using genetically modified plants with smaller PSU sizes and higher photosynthetic conversion efficiencies will still be at the mercy of extrinsic factors and weather uncertainties.

In our opinion the ultimate agriculture of the future should consist of bioreactors populated with bioengineered, highly efficient photosynthetic membranes, with a small PSU size and operating at efficiencies that approach the 12% maximal theoretical efficiency of the PETS that may be observed under white light, or the 27% maximal theoretical efficiency that may be achieved under red light. Such conditions may be set up during space travel or in large space stations (25). The photosynthetic product may well be a short chain carbohydrate such as glycerol that can be converted into food fiber and energy. In the meanwhile, let us not forget that a journey of 10,000 miles starts with the first step.

Abbreviations

ALA, δ-aminolevulinic acid; Chl, chlorophyll; Chlide, chlorophyllide; Pchlide, protochlorophyllide; Pchl(ide), Pchlide and/or Pchlide ester; Proto,

**Table II. Calculated distances R, Separating Proto, Mp(e) and Pchlide *a*
Donors from Chl *a*-Protein Complexes Acceptors in Barley and Cucumber
Chloroplasts at 77°K *in Situ***

Chl *a*	Proto		Mp(e)		MV Pchlide *a*	DV Pchlide *a*
Species	B	C	B	C	B	C
			R (Å)			
Chl *a* F685 (LHCI-680 + outer half of LHCII)	44.1	39.0	39.7	38.2	37.3	34.5
Chl *a* F695 (CP47) + CP29)	48.3	42.7	43.5	41.8	40.8	37.0
Chl *a* F735 (LHCI-730)	14.6	12.9	13.1	12.6	12.3	11.2

NOTE: The distances R, were determined from the following calculated
parameters: (a) the critical separation R_0 of donors and acceptors, (b) the
efficiency E of resonance excitation energy transfer from donors to Chl*a*
acceptors. The overlap integral needed for the calculation of R_0 was
approximated using a derived formula instead of numerical integration. B =
barley, C = cucumber.
SOURCE: Adapted from ref (24).

protoporphyrin IX; Mp, Mg-Proto; Mpe, Mg-Proto monomethyl ester; Mp(e),
Mp and/or Mpe; DV, divinyl; MV, monovinyl; LHC, light harvesting Chl-
protein complex; PETS, photosynthetic electron transport system; PS,
photosystem; PSI, photosystem I; PSII, photosystem II; PSU, photosynthetic
unit; RC, reaction center; LHCII, the major thylakoid, LHC outer antenna;
LHCI-680, an LHC antenna of PSI; LHCI-730, another LHC antenna of PSI;
CP47, a core antenna of PSII; CP29, an inner LHC antenna of PSII; Å,
angstrom; s, second; h, hour.

104

Note: Unless proceeded by MV or DV tetrapyrroles are used generically to designate metabolic pools that may consist of MV and/or DV components.

Acknowledgments

This work was supported by funds from the Illinois Agricultural Experiment Station, by the John P. Trebellas Photobiotechnology Research Endowment, and by the C.A. and C. C. Rebeiz Endowment for Basic Research.

References

1. Lien, S.; San Pietro, A. "An inquiry into the biophotolysis of water to produce hydrogen,"RANN Report, Indiana University, 1975.
2. Rebeiz, C. A.; Kolossov, V. L.; Briskin, D.; Gawienowski, M. Chloroplast Biogenesis: Chlorophyll biosynthetic heterogeneity , multiple biosynthetic routes and biological spin-offs. In *Handbook of Photochemsistry and Photobiology*; N. S. Nawla, Ed.; American Scientific Publishers: Loseles, CA., 2002; pp 183-268.
3. Rebeiz, C. A.; Parham, R.; Fasoula, D. A.; Ioannides, I. M. Chlorophyll biosynthetic heterogeneity. In *The Biosynthesis of the Terapyrrole Pigments*; D. J. Chadwick and K. Ackrill, Eds.; John Wiley and Sons: New York, 1994; pp 177-193.
4. Rebeiz, C. A. Analysis of intermediates and end products of the chlorophyll biosynthetic pathway. In *Heme Chlorophyll and Bilins, Methods and Protocols*; M. Witty, Ed.; Humana Press: Totowa NJ, 2002; pp 111-155.
5. Smith, J. H. C.; French, C. S. The major accessory pigment in photosynthesis. *Ann. Rev. Plant Physiol.* **1963**, *14*, 181-224.
6. Rebeiz, C. A.; Castelfranco, P. Protochlorophyll biosynthesis in a cell-free system from higher plants.*Plant Physiol.* **1971**, *47*, 24-32.
7. Rebeiz, C. A.; Castelfranco, P. Chlorophyll biosynthesis in a cell-free system from higher plants. *Plant Physiol.* **1971**, *(47)*, 33-37.
8. Daniell, H.; Rebeiz, C. A. Chloroplast Culture VIII. A new effect of kinetin in enhancing the synthesis and accumulation of protochlorophyllide *in vitro. Biochem. Biophys. Res. Commun.* **1982**, *104*, 837-843.
9. Daniell, H.; Rebeiz, C. A. Chloroplast Culture IX. Chlorophyll(ide) *a* biosynthesis in vitro at rates higher than in vivo. *Biochem. Biophys. Res. Commun.* **1982**, *106*, 466-470.
10. Rebeiz, C. A.; Montazer-Zouhoor, A.; Daniell, H. Chloroplast Culture X: Thylakoid assembly *in vitro. Isr. J. Bot.* **1984**, *33*, 225-235.

11. Kolossov, V.; Ioannides, I. M.; Kulur, S.; Rebeiz, C. A. Chloroplast biogenesis 82: Development of a cell-free system capable of the net synthesis of chlorophyll(ide) *b*. *Photosynthetica* **1999**, *36*, 253-258.

12. Carey, E. E.; Rebeiz, C. A. Chloroplast Biogenesis 49. Difference among angiosperms in the biosynthesis and accumulation of monovinyl and divinyl protochlorophyllide during photoperiodic greening. *Plant Physiol.* **1985**, *79*, 1-6.

13. Ioannides, I. M.; Fasoula, D. M.; R., R. K.; Rebeiz, C. A. An evolutionary study of chlorophyll biosynthetic heterogeneity in green plants. *Biochem. Sys. Ecol.* **1994**, *22*, 211-220.

14. Abd-El-Mageed, H. A.; El Sahhar, K. F.; Robertson, K. R.; Parham, R.; Rebeiz, C. A. Chloroplast Biogenesis 77. Two novel monovinyl and divinyl light-dark greening groups of plants and their relationship to the chlorophyll *a* biosynthetic heterogeneity of green plants. *Photochem. Photobiol.* **1997**, *66*, 89-96.

15. Rebeiz, C. A.; Ioannides, I. M.; Kolossov, V.; Kopetz, K. J. Chloroplast Biogenesis 80. Proposal of a unified multibranched chlorophyll *a/b* biosynthetic pathway. *Photosynthetica* **1999**, *36*, 117-128.

16. *Pigment-Protein Complexes in Plastids: Synthesis and Assembly*; Sundqvist, C.; Ryberg, M., Eds.; Academic Press: New York, 1993.

17. Bassi, R.; Rigoni., F.; Giacometti, G. M. Chlorophyll binding proteins with antenna function in higher plants and green algae. *Photochem. Photobiol.* **1990**, *52*, 1187-1206.

18. Kolossov, V. L., Kopetz, K. J., Rebeiz, C. A. Chloroplast Biogenesis 87: Evidence of resonance excitation energy transfer between tetrapyrrole intermediates of the chlorophyll biosynthetic pathway and chlorophyll *a*. *Photochem. Photobiol.* **2003**, *78*, 184-196

19. Calvert, J. G.; Pitts, J. N. *Photochemistry*; John Wiley & Sons: New York, 1967; 899.

20. Lakowicz, J. R. *Principles of Fluorescence Spectroscopy*; Kluwer Academic/Plenum Press: New York, 1999; 698.

21. Turro, N. J. *Molecular Photochemistry*; W. A. Benjamin: London, 1965; 286.

22. Rebeiz, C. A.; Montazer-Zouhoor, A.; Mayasich, J. M.; Tripathy, B. C.; Wu, S. M.; Rebeiz, C. C. Photodynamic Herbicides. Recent developments and molecular basis of selectivity. *Crit. Rev. Plant Sci.* **1988**, *6*, 385-434.

23. Jones, O. T. G. Magnesium 2,4-divinyl phaeoporphyrin a₅ monomethyl ester, a protochlorophyll-like pigment produced by *Rhodopseudomonas spheroides*. *Biochem. J.* **1963**, *89*, 182-189.

24. Kopetz, K. J. Topography of the Multibranched Integrated Chlorophyll *a/b* Biosynthetic Pathway, MS Thesis, University of Illinois, 2000.

25. Rebeiz, C. A.; Daniell, H.; Mattheis, J. R. Chloroplast Bioengineering: The greening of chloroplasts in vitro. In *In :Biotech. Bioeng. Symp. 12, C. D. Scott, ed*, 1982; pp 414-439.

Chapter 8

Subplastidic Distribution of Chlorophyll Biosynthetic Intermediates and Characterization of Protochlorophyllide Oxidoreductase C

Baishnab C. Tripathy[*], Anasuya Mohapatra,
and Gopal K. Pattanayak

School of Life Sciences, Jawaharlal Nehru University,
New Delhi 110067, India
[*]Corresponding author: bctripathy@mail.jnu.ac.in

Chloroplasts isolated from leaves of *Beta vulgaris* were
fractionated to thylakoids, stroma, outer and inner
envelope membranes. In addition to protochlorophyllide,
both outer and inner envelope membranes contained
protoporphyrin IX and Mg-protoporphyrin IX + its
monoester. However, the pigment content of inner
membrane was higher than that of the outer. The
proportion of monovinyl and divinyl forms of protochlorophyllide was similar in intact plastid, thylakoids,
outer and inner envelope membranes suggesting a tight
regulation of vinyl reductase enzyme. The significance of
differential distribution of tetrapyrroles among thylakoids,
outer and inner envelope membranes and stroma is
discussed in relation to the distribution of
protochlorophyllide oxidoreductase. The presence of a
new isoform of a protochlorophyllide oxidoreductase C
gene that codes for the protein that phototransforms
protochlorophyllide to chlorophyllide is demonstrated.

Chlorophyll (Chl) is essential to photosynthesis and crop productivity. Chl biosynthesis produces the greening of plants in spring and its degradation is manifested in the loss of green pigmentation in fruit ripening and autumnal deciduous leaves in temperate zones. Chlorophyll is bound to pigment-protein complexes of thylakoid membranes. Chl and its precursors are essential for chloroplast development and nuclear gene expression *(1,2,3,4)*. Protochlorophyllide (Pchilde), an intermediate of Chl biosynthetic pathway, is present both in the envelope and thylakoid membranes *(5)*. 5-Aminolevulinic acid (ALA) biosynthetic enzymes are located in the stroma *(6)*. Interplay of envelope, stroma and thylakoids is shown for protoporphyrin IX (Proto IX) synthesis and enzymes responsible for conversion of ALA to Proto IX i.e., ALA dehydratase, porphobilinogen deaminase, uroporphyrinogen decarboxylase and coproporphyrinogen oxidase are mostly located in the stromal phase *(6)*. All subsequent steps of Chl biosynthesis are catalyzed by membrane bound or membrane associated enzymes *(7,8,9,10)*. Therefore, three subplastidic compartments interact during tetrapyrrole synthesis. However, to understand the coordinated interaction of different subplastidic compartments to synthesize Chl and to visualize the intraplastidic route of Chl biosynthesis it is essential to know the distribution of Chl biosynthetic intermediates among different plastidic compartments i.e., envelope, stroma and thylakoids. Although protochlorophyllide (Pchlide) is present both in the thylakoid and envelope membranes *(5)* sub-plastidic localization of other important Chl biosynthetic intermediates i.e., Proto IX and Mg-protoporphyrin IX + its monoester (MP(E)) has not been studied. Besides quantitative distribution of Pchlide, MP(E) and Proto IX among different sub-plastidic compartments is not known. The distribution of pigments in the outer and inner envelope membrane is not studied.

In this article we present a detailed report of the localization and relative distribution of Proto IX, MP(E) and Pchlide, tetrapyrrolic intermediates of Chl biosynthetic reactions, among different subplastidic compartments i.e. envelope, stroma and thylakoids. Besides the envelope membrane is further sub-fractionated to outer and inner components and distribution of these pigments in outer and inner envelope membranes is estimated. It is shown that in addition to Pchlide, both outer and inner envelope membranes contain Proto IX and MP(E). The significance of differential quantitative distribution of chlorophyll biosynthetic intermediates among thylakoids, envelope and stroma and the presence of chlorophyll biosynthetic intermediates in the outer envelope membrane is discussed in relation to Chl biosynthetic route inside the plastid and the presence of protochlorophyllide oxidoreductase. Two protchlorophyllide oxidoreductase (POR) genes i.e., *por A* and *por B* are identified and their functions are well established in higher plants *(11,12,13)*. Based on sequence homology with that of *por A* and *por B* genes, a novel *por C* gene was identified from *Arabidopsis thaliana (14)*. However, its function is not demonstrated. In the present investigation we have shown its catalytic function. The mature *por C* devoid of transit peptide is overexpressed in a heterologous system *Escherichia coli.* It is demonstrated that the overexpressed por C protein reduces Pchlide to Chlide in a light-dependent manner.

MATERIALS AND METHODS

Plant Material

Beta vulgaris L. was grown in the garden of Jawaharlal Nehru University, New Delhi.

Treatment with 5-aminolevulinic acid and α,α'-dipyridyl

Leaves were harvested and 10 g batches of leaves were incubated in 40 ml solution of 4 mM ALA and 4 mM α,α'-dipyridyl in dark to accumulate porphyrins *(15)*.

Isolation of Purified Intact Chloroplasts

Intact chloroplasts were isolated from 60 g batches of leaves over a percol gredient as described before *(16,17)* .

Preparation of Purified Stroma, Envelope and Thylakoid Membranes

In order to fractionate lysed chloroplast into stroma, envelope and thylakoid, chloroplasts were lysed in ice bucket by suspending them for 10 min in TE buffer consisting of 10 mM Tris-HCl, pH 7.5 and 1 mM EDTA. Lysed chloroplasts were centrifuged in a sucrose step gradient in Sorvall, Combi plus or Beckman, XL-90 ultracentrifuge at 40,000 rpm in TH 641/SW 41 swinging bucket rotor for 1h with slow acceleration without brake (6). After centrifugation envelope membranes appeared as yellow band at the interface of 0.46 M and 1.0 M sucrose gradient. Stroma remained at the top of the gradient whereas the thylakoid membranes settled at the bottom *(18)*. All operations were done in diffused green light to minimize phototransformation of Pchlide to Chlide.

Envelope membranes, collected from the interface of 0.46 M and 1.0 M sucrose gradient, were washed once with 10 ml of TE buffer and centrifuged in SW 41 rotor at 40,000 rpm for 30 min to pellet the envelope membranes. The envelope membranes were resuspended in 200 µl of TE buffer and reloaded on the sucrose step gradient and centrifuged in Sorvall, Combi plus or Beckman, XL-90 ultracentrifuge at 40,000 rpm in TH 641/SW 41 swinging bucket rotor for 1h with slow acceleration without brake to remove contaminants. After centrifugation the yellow band of envelope membranes at the interface of 0.46 M and 1.0 M sucrose gradient was c ollected. It was washed with 10 ml of TE buffer and centrifuged in SW 41 rotor at 40,000 rpm for 30 min to pellet the envelope membranes.

Thylakoid membranes were washed 5 times in the buffer consisting of 10 mM Tris (pH 7.5) and 0.4 M sucrose to remove contaminating envelope membranes. While doing quantitative estimation of tetrapyrroles in different fractions of plastid, supernatant fractions collected from 5 washings of thylakoid membranes were combined, diluted 5 fold in TE and centrifuged at 40,000g for 20 minutes to pellet the membranes. They were resuspended in 0.3 ml of TE buffer and were subsequently layered on top of the discontinuous sucrose gradient and centrifuged in a Sorvall, Combi plus or Beckman, XL-90 ultracentrifuge at 40,000 rpm in TH 641/SW 41 swinging bucket rotor for 1 h w ith s low a cceleration w ithout b rake. A fter centrifugation envelope membranes at the interface of 0.46 M and 1.0 M sucrose gradient were collected and washed once with 10 ml of TE buffer and centrifuged in SW 41 rotor at 40,000 rpm for 30 min to pellet the envelope membranes. The envelope membranes were resuspended in minimal amount of TE buffer. The envelope membrane fractions obtained from both centrifugations were combined and their tetrapyrrole contents were estimated. A minute amount of thylakoid membrane which formed a pellet at the bottom of the centrifuge tube was collected and combined with the main thylakoid membrane fraction.

The stroma that remained at the top of the gradient was collected along with the entire amount of 0.46 M sucrose immediately above the interface of envelope membranes. It was diluted to 3 volumes in TE buffer and was centrifuged in Beckman SW 50.1 rotor at 1,50,000g for 2 h to remove contaminating membrane vesicles. All operations were done at 4 C.

Preparation of Outer and Inner Envelope Membranes

Envelope m embranes were prepared by slight modification of the method of (18,19). Intact chloroplasts were suspended in hypertonic buffer consisting of 1.3 M sucrose, 2 mM EDTA and 10 mM Tricine-NaOH buffer pH 7.5 at a concentration of 40 mg Chl. Chloroplasts were given freeze and thawing shocks twice and the volume was adjusted to 21 ml with the hypertonic lysis buffer. The broken 21 ml chloroplast suspension was transferred to a 35 ml centrifuge tube and 8 ml of 1.2 M, 3 ml of 1.1 M and 3 ml of 0.2 M sucrose prepared in 10 mM Tricine-NaOH buffer pH 7.5, 2 mM EDTA and 5 mM $MgCl_2$ were successively gently layered over the top of the chloroplast suspension. The gradient was centrifuged at 1,22,000g for 15 h at 4 C. The envelope membranes were isolated by a bove f loatation centrifugation and were collected as yellow band at the interface between 0.2 M and 1.1 M sucrose layer. The envelope membranes were layered on a linear sucrose gradient from 0.3 M to 1.1 M sucrose containing 10 mM Tricine-NaOH buffer pH 7.5, 2 mM EDTA and 5 mM $MgCl_2$ and were centrifuged at 1,22,000g for 15 h at 4°C to separate the outer and inner envelope membranes. One ml fractions were collected by a peristaltic pump and a fraction collector and their absorbance was measured at 280 nm and 678 nm.

Checking Purity of Intact Chloroplast, Envelope, Stroma and Thylakoid Fractions

To test the purity of intact chloroplasts, envelope membranes, thylakoid membranes and stromal preparations certain marker enzymes or proteins were monitored as follows.

Succinate cytochrome c reductase

The mitrochondrial marker enzyme, succinate cytochrome c reductase reduces cytochrome c utilizing electrons donated by succinate. Instead of cytochrome c, 2,6-dichlorophenol indophenol was used as an electron acceptor (20).

Gluconate-6-phosphate dehydrogenase

The stromal marker enzyme, gluconate-6-phosphate dehydrogenase, converts gluconate-6-phosphate to D-ribulose 5-phosphate in an oxidative decarboxylation reaction and simultaneously reduces $NADP^+$ to NADPH. The enzymatic activity was calculated as nmol of $NADP^+$ reduced mg protein^{-1} min^{-1} using absorption coefficient, 6.22×10^3 M^{-1}cm^{-1} (10,21).

Mg^{2+}-ATPase

The activity of the marker enzyme for envelope membranes, Mg^{2+}-ATPase was monitored by adding 0.1 ml of sample to be assayed for envelope contamination to 0.9 ml of reaction mixture (pH 8.0) consisting of 20 mM Tricine-NaOH, 1 mM $MgCl_2$, 1 mM ATP, 0.2 mM NADH, 0.5 mM phosphoenolpyruvate, 1.5-3.0 units of pyruvate kinase and 2-3 units of lactate dehydrogenase. Activity was calculated using absorption coefficient for NAD, 6.22×10^3 M^{-1} cm^{-1} (22).

Immunoblotting of Envelope and Thylakoid Membranes

Forty μg protein of envelope or thylakoid membranes were loaded for SDS-PAGE and were transferred to a nitrocellulose membrane. Western blotting of outer membrane protein 14 (OM 14) *(30)* (OM 14 antibody was a gift of Professor Kenneth Keegstra, East Lansing, USA) and LHCP II (LHCP II was a gift of Professor A. K. Matto, Beltsville, USA) were performed in a Bio-Rad Transblot apparatus using anti-rabbit secondary antibody coupled to alkaline phosphatase *(2)*.

Separation of Pigments

Pigments from envelope or thylakoid membranes were extracted into 80% acetone, centrifuged at 4 ^{0}C in a microfuge for 5 min at maximum speed. Hexane-extracted acetone residue solvent mixture (HEAR) was prepared from 80% acetone e xtract b y a dding h exane a s d escribed b efore *(23)*. HEAR contained chlorophyllide (Chlide), protochlorophyllide (Pchlide), protoporphyrin IX (Proto IX) and other non-esterified porphyrins.

In order to take low temperature spectra (77 K) of the pigments present in HEAR, they were transferred to ether phase as follows. To the HEAR, 1/10 volume of saturated NaCl, 1/30 volume of 0.5 M phosphate buffer (pH 7.0) and 1/4 volume of diethyl ether were added and mixed thoroughly in a separating funnel and allowed to stand and the upper ether layer was collected. The ether phase collected was washed with 0.5 M phosphate buffer (pH 7.0) thrice and was used for fluorometric studies.

Spectrofluorometry

Fluorometric estimation of pigments was done using SLM Aminco 8000 photon counting spectrofluorometer. The channel A (sample) and channel C (reference) were adjusted to 20,000 counts per second using tetraphenylene butadiene block as standard, excited at 348 nm and fluorescence emitted monitored at 422 nm. The samples were excited at appropriate wavelength and emission spectra were recorded in ratio mode (channel A/C) at excitation and emission slit widths of 4 nm. Spectra were c orrected f or photomultiplier tube response. Excitation spectra were recorded in ratio mode (channel A/C) having slit width of excitation monochromator set at 4 nm and that of emission monochromator set at 8 nm. Using appropriate equations concentrations of Proto IX, MPE , MV Pchlide and DV Pchlide were quantified *(24,25)*. Low temperature spectra were taken by freezing the sample at 77K using liquid nitrogen in a Dewar cuvette. MV and DV forms of Pchlide were quantified from their 77K fluorescence excitation spectra (F625) *(26)*.

Chlorophyll and Protein Estimation

Chl and protein were estimated according to *(27)* and *(28)* respectively.

Protein Expression and Assay of POR C Enzymatic Activity

Nucleotide sequences encoding the mature POR C protein (lacking the codons for the first 67 amino acids belonging to the transit peptide) of *A. thaliana* *(14)* were amplified by PCR from cloned pGEM T-easy (Promega, USA) plasmids containing the full length *por* C cDNA *(29)* The gene was overexpressed in pET 28a expression vector having hexahistidine affinity tag and transformed into *E.coli* BL 21 (DE3) cells which were induced with 0.2 mM isopropyl β-D-thiogalactoside (IPTG) at 30^0C for 3 h to express POR C. Proteins were purified using Ni-NTA resin column (Qiagen, USA) having the molecular weight of 40.15 kDa that includes the 36.85 kDa mature POR C protein..

Assay of POR C Enzymatic Activity: POR C enzymatic activity was assayed in reaction buffer consisting of 50 mM HEPES/KOH, pH 7.5, 2mM $MgCl_2$, 0.05% Triton X-100 (w/v), supernatant fraction of *E. coli* cell lysate having overexpressed protein, 100 nM Pchlide, and 5 mM NADPH. The reaction mixture was allowed to equilibrate at room temperature (25^0C) for 15 minutes in complete darkness. After 15 minutes of dark incubation the reaction mixture was illuminated with cool white fluorescent light (20 μmol m^{-2} s^{-1}) for 2, 15, 30, 60 and 120 minutes respectively and their fluorescence spectra were recorded. The samples were excited at 433 nm.

Chemicals

All porphyrin intermediates were purchased from Porphyrin Products, Logan, UT. Other chemicals were purchased from Sigma chemical Co., Merck, Sd fine chemicals, BDH and Qualigens.

RESULTS

Purity of Chloroplast Preparation

Intact chloroplasts isolated over a percoll gradient were assayed for mitochondrial contamination by monitoring the activity of succinate cytochrome c reductase *(20)*, an oxidative electron transport chain component of mitochondria as the marker enzyme. There was negligible activity of this enzyme in the intact chloroplast preparation (data not shown) suggesting that the intact chloroplast preparation was free from mitochondrial contamination.

Purity of Stroma, Envelope and Thylakoid Fractions

Intact chloroplasts were lysed by osmotic shocks and separated into stroma, envelope and thylakoid fractions. To check the purity of stromal preparation the Mg^{2+}-ATPase, present in chloroplast envelope membranes, was used as the marker enzyme to determine the envelope membrane contamination *(10,22)*. Similarly, the purity of envelope membrane was monitored by measuring the activity of the stromal marker enzyme gluconate-6-phosphate dehydrogenase *(10,21)*. The Mg^{2+}-ATPase activity was highest in the envelope membrane fraction and was minimal in the stroma (Table-I).

Gluconate-6-phosphate dehydrogenase activity was present in the stromal fraction and was absent in the envelope membrane. This demonstrated that envelope membrane preparation was almost free from stromal contamination (Table-I). Thylakoids did not have significant gluconate-6-phosphate dehydrogenase activity which demonstrated that they were free from stromal contamination. Chl was used as a marker pigment to detect thylakoid contamination. The Chl(ide) content of chloroplast envelope was 4 ± 1 µg (mg protein)$^{-1}$. This value closely matched with that of previous report *(5)* suggesting that envelope was almost free from thylakoids. The stroma was almost free of thylakoids as it did not have significant amounts of Chl.

To study cross contamination of thylakoid and envelope membranes the outer envelope membrane marker protein OM 14 and thylakoid membrane marker protein light harvesting chlorophyll protein complex II (LHCPII) were immunoblotted. The immunoblot of envelope and washed thylakoid membrane proteins with OM 14 antibodies *(30)* showed that it was present in the envelope and absent in the thylakoid membranes which demonstrated the purity of thylakoid membranes (Fig.1A). The immunoblot of envelope and thylakoid membrane proteins with LHCPII antibodies showed a small feeble band in the envelope and a strong band in the thylakoid membrane (Fig.1B). Envelope and thylakoids were loaded on equal protein basis (50 µg). Per chloroplast, the amount of envelope protein is much lower than that of thylakoids. Therefore, as compared to thylakoids, 50 µg envelope protein had to be isolated from a larger number of plastids i.e. it was an enrichment process for envelope membrane. These demonstrated that envelope membranes were almost free from thylakoid contamination.

Table I. Gluconate-6-phosphate dehydrogenase and Mg^{2+} ATPase activities and Chl content in stroma, envelope and thylakoid fractions

----

Sample	Gluconate-6-phosphate dehydrogenase	Mg^{2+}-ATPase	Chl

----

	nmol NADPH	*nmol*	*μg (mg protein)⁻*
	mg protein⁻¹ min⁻¹	*Pi mg protein⁻¹ min⁻¹*	
Stroma	120.4 ± 8.0	5.4 ± 1.0	0.06±0.01
Envelope	0	219.9 ± 9.8	4.0±1.0
Thylakoids	2.0± 0.2	not determined	160.0±10.0

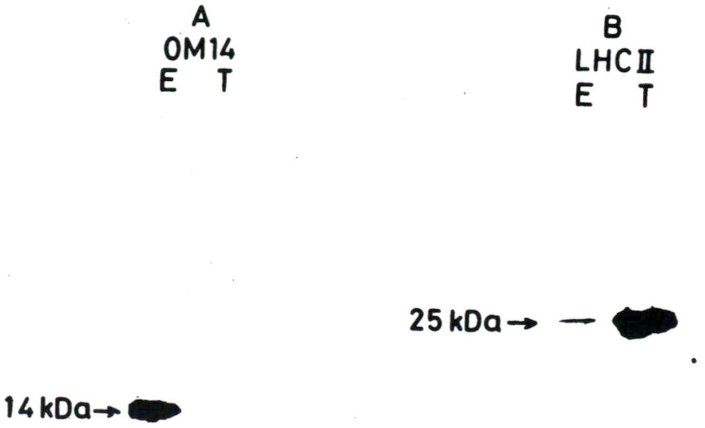

A
OM14
E   T

B
LHC II
E   T

25 kDa → — ▰

14 kDa→ ▰

Figure 1. Western blot of envelope (E) and thylakoid (T) membrane proteins probed with antibodies of (A) OM 14 and (B) LHCPII.

Pigment Contents of Envelope, Stroma and Thylakoid Fractions

Before isolation of chloroplasts, concentration of Chl biosynthetic intermediates were monitored from leaf samples. Per g fresh weight, leaves had 0.14 nmol of Proto IX, 0.82 nmol of MP(E) and 9.8 nmol of Pchlide. These concentrations match with those determined from other plant species (31). Table-II shows the quantitative distribution pattern of Chl biosynthetic intermediates among subplastidic fractions of envelope, stroma and thylakoids.

Table II. Subplastidic partitioning of Chl biosynthetic intermediates to stroma, envelope and thylakoid fraction

Subplastidic fraction	Proto IX	MP(E)	Pchlide
	(n mol)		
	Pigments loaded on top of the gradient		
	0.057	0.34	3.97
	Pigments recovered from different fractions of the gradient		
Stroma	traces	traces	traces
Envelope	traces	traces	0.039±0.004
Thylakoid	0.05±0.01	0.30±0.03	3.50±0.2
Total recovery	0.05	0.30	3.54
Recovery loss (%)	12.28	11.76	10.83

Analysis of these pigments were done by a photon counting spectrofluorometer and the accuracy of quantitation was as good as that done by HPLC (23,24). There were very small amounts of Proto IX and MP(E) in different subplastidic preparations. However, substantial amount of non-phototransformable Pchlide was present in thylakoids. The Pchlide content of envelope membranes was 1.1% of total plastidic Pchlide.

To understand the subplastidic localization and distribution of Proto IX and MP(E), two other important intermediates of Chl biosynthetic pathway, it was necessary to poise the green leaves in Chl biosynthetic mode. This was achieved by incubating leaves in dark with ALA, the precursor of tetrapyrroles, and the Fe-chelator α,α'-dipyridyl. This induced green leaves to synthesize and accumulate Proto IX, MP(E) and Pchlide (31). Fe-chelator α,α'-dipyridyl is an inhibitor of MPE cyclase and partially inhibits Pchlide synthesis (32). However, when applied in conjunction with ALA, they

induce a ccumulation o f m assive a mounts of Proto IX, MP(E) and Pchlide
(31). One g of leaves of *Beta vulgaris* treated with ALA + α,α'-dipyridyl
accumulated 30.21 nmol Pchlide, 4.26 nmol MPE and 3.05 nmol Proto IX.
Under identical conditions, ALA alone induced accumulation of 99.0 nmol
Pchlide, 1.0 nmol of MP(E) and 14.0 nmol Proto IX.

As its treatment resulted in the synthesis of substantial amounts of
MP(E) in addition to Proto IX and Pchlide, intact chloroplasts isolated from
ALA + α,α'-dipyridyl-treated leaves were fractionated to stroma, outer and
inner envelope membranes and thylakoid membranes and their porphyrin
contents were estimated. All operations were done in diffused green light to
minimize phototransformation of Pchlide to Chlide. The total recovery of
porphyrins from d ifferent fractions in the sucrose g radient was 8 3-90% of
that of total plastidic lysate loaded on the top of the gradient. As shown in
Table-III, plastidic Proto IX was partitioned to stroma, envelope and
thylakoid membranes. Proto IX partitioned maximally to thylakoids
(65.57%) followed by stroma (33.77%) and envelope (0.66%). MP(E)
content in different subplastidic fractions were 89.0% for thylakoids, 10.0%
for stroma and 1.0% for envelope. Pchlide partitioned maximally to
thylakoids (98.48%), followed by envelope (1.5%) and to a negligible extent
(0.02%) into stroma.

**Table III. Subplastidic distribution of Chl biosynthetic intermediates
among stroma, envelope and thylakoid fractions of intact chloroplasts
isolated from leaves treated with ALA+ α,α'-dipyridyl**

Subplastidic fraction	Proto IX	MP(E)	Pchlide
	(n mol)		
	Pigments loaded on top of the gradient		
	3.65	10.00	71.14
	Pigments recovered from different fractions of the gradient		
Stroma	1.03±0.14 (33.77)	0.89±0.1 (10)	0.01±0.002 (0.02)
Envelope	0.02±0.005 (0.66)	0.09±0.01 (1)	0.94±0.11 (1.5)
Thylakoid	2.00±0.2 (65.57)	7.89±0.9 (89)	61.46±6.0 (98.48)
Total recovery	3.05	8.87	62.41
Recovery loss (%)	16.44	11.30	12.27

Since envelope membranes contain only small amounts of tetrapyrroles, contamination of stroma with trace amounts of envelope (Table-I) cannot account for large amounts of Proto IX or MP(E) in the stroma. As trace amounts of thylakoids were present in the envelope fraction (Fig.1) it was important to probe if Chl biosynthetic intermediates estimated from envelope fraction was due to thylakoid contamination. For the calculation purpose it was assumed that all the Chl(ide) estimated from envelope membrane fraction was due to thylakoid contamination (this assumption is not true). Amounts of Pchlide, MP(E) and Proto IX present in the thylakoid membranes were expressed as nmols (mg Chl)$^{-1}$. Using these values, Pchlide, MP(E) and Proto IX were calculated as nmols per mg of Chl(ide) present in the envelope membrane fraction. The calculated amounts of Pchlide, MP(E) and Proto IX were only 5-15% of individual pigments actually estimated from the envelope fraction (data not shown). This clearly demonstrates the presence of Chl biosynthetic intermediates i.e., Pchlide, MP(E) and Proto IX in the envelope membrane and disproves the unfounded speculation that they were due to contaminating thylakoids.

Distribution of Pigments in Outer and Inner Envelope Membranes

To further understand the distribution of Chl biosynthetic intermediates among outer and inner envelope membranes, their tetrapyrrole contents were estimated by spectrofluorometry. Several intermediate membrane fractions collected from sucrose gradient were mixtures of both outer and inner envelope membranes and were discarded. Usually there is slight contamination of outer envelope in the inner envelope membrane preparation due to presence of contact sites *(18)*. Therefore, it was not possible to quantify the absolute amounts of pigments present in outer and inner envelope membrane. The pigments from purified outer and inner membrane fractions were estimated and their pigment content was expressed as nmol (mg protein)$^{-1}$. The distribution of pigments between inner and outer envelope membranes is shown in Table-IV.

All the pigments i.e., Proto IX, MP(E), Pchlide and Chlide are present both in outer and inner envelope membranes. However, pigment contents of outer membrane are invariably smaller than those of inner membranes.

Characterization of Tetrapyrroles

To ascertain the physiological status of Pchlide-POR ternary complex present in outer and inner envelope membranes, their low temperature emission spectra were recorded. As shown in Figure 2, the 77K fluorescence emission spectra (E440) of both outer and inner membranes suspended in 10 mM TE buffer exhibited a peak at 678 nm due to Chl(ide) and at 632 nm due to non-phototransformable Pchlide. Envelope membranes were isolated from light-grown green plants. Consequently, both outer and inner envelope membranes lack the peak at 656 nm due to

Table IV. Distribution of Chl biosynthetic intermediates between the outer and inner envelope membranes

Membrane fraction	Proto IX	MP(E)	Pchlide	Chl(ide)
		nmol (mg protein)$^{-1}$		
Outer Membrane	0.03±0.01	0.11±0.01	1.25±0.14	1.0±0.11
Inner Membrane	0.07±0.01	0.27±0.03	2.65±0.30	3.5±0.40

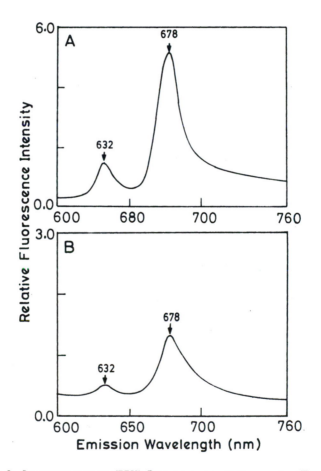

Figure 2. Low temperature (77K) fluorescence emission spectra (E440) of inner (A) and outer (B) envelope membranes suspended in 10 mM TE buffer.

phototransformable Pchlide. Most of the Pchlide synthesized in response to ALA treatment mostly accumulate in non-phototransformable form *(33)*.

Pchlide in plants is present both in monovinyl and divinyl forms *(34,35,36)*. To understand if there is any differential distribution of monovinyl Pchlide (MV Pchlide) and divinyl Pchlide (DV Pchlide) in thylakoids, inner and outer envelope membranes, pigments from these fractions were extracted and transferred to ether layer and their low temperature (77K) fluorescence spectra were recorded. Figure 3 shows the 77K fluorescence emission spectra (E440) of ether extracts of pigments isolated from (A) intact chloroplasts , (B) thylakoids , (C) inner and (D) outer envelope membranes. They had fluorescence peaks at 625 nm due to Pchlide and at 674 nm due to Chl(ide). Both MV and DV Pchlide have the fluorescence emission maximum at 625 nm. Therefore, they could not be resolved from their fluorescence emission maxima. However, they could be resolved on the basis of their fluorescence excitation peaks. M V P chlide exhibits a fluorescence excitation (F625) peak at 437 nm, and a hump at 443 nm. DV Pchlide has a fluorescence excitation (F625) peak at 443 nm and a hump at 451 nm. F igure 4 shows the 77K fluorescence excitation spectra (F625) of Pchlide extracted from (A) intact chloroplasts , (B) thylakoids , (C) inner and (D) outer envelope membranes. All the samples had peaks at 437 nm and a hump at 443 nm predominantly due to MV Pchlide. The hump at 451 nm was very small. Applying appropriate equation the proportion of MV and DV forms of Pchlide was estimated *(26)*. As shown in Table-V, nearly 90% of Pchlide was in monovinyl form and around 10% of them was in divinyl form in plastidic and subplastidic preparations i.e., chloroplasts, thylakoids, outer and inner envelope membranes. This d emonstrates that t here i s n o d ifferential d istribution of MV Pchlide and DV Pchlide in different subplastidic compartments.

Protochlorophyllide Oxidoreductase C Activity

To d etermine i f POR C p rotein p hototransforms P chlide to Chlide its enzymatic activity was monitored. The POR C enzyme was i ncubated i n dark with NADPH and Pchlide. At room temperature, when it was excited at 433 nm (E433) had a fluorescence emission peak at 636 nm (Figure 5). After illumination with cool white fluorescent light (20 μmol m^{-2} s^{-1}) for different l engths o f t ime, a p eak a t 6 73 n m d ue t o C hlide a ppeared. The fluorescence emission peak at 673 nm was visible after 2 min of illumination. The peak height at 673 nm increased with concomitant decrease of 636 nm emission after 15 minutes, 30 minutes, 1 h and 2 h of illumination demonstrating the gradual conversion of Pchlide to Chlide. Uninduced bacterial lysate in presence of NADPH and Pchlide and induced

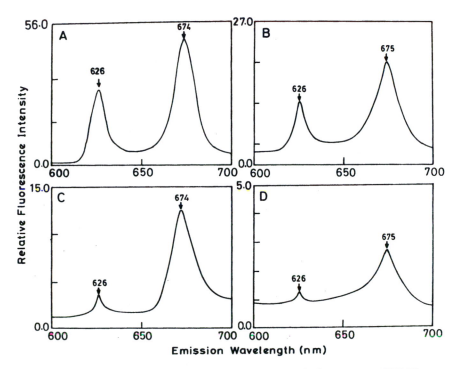

Figure 3. Fluorescence (77 K) emission spectra of ether extract of HEAR fraction of pigments extracted from intact chloroplasts (A), thylakoid membranes (B), inner envelope membrane (C) and outer envelope membrane (D). Samples were excited at 440 nm having slit widths of 4 nm.

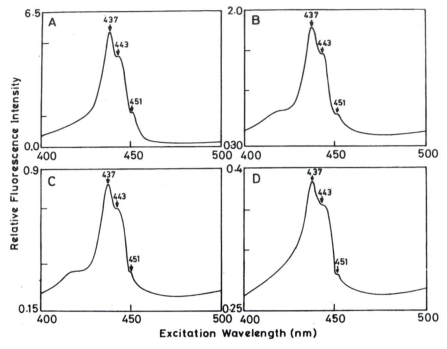

Figure 4. Fluorescence excitation spectra recorded at 77 K of ether extract of HEAR fraction of pigments extracted from intact chloroplasts (A), thylakoid membranes (B), inner envelope membrane (C) and outer envelope membrane (D). Emission wave length was fixed at 625 nm and excitation spectrum was recorded having excitation and emission slit widths of 4 nm.

Table-V. Distribution of monovinyl and divinyl Pchlide in different subplastidic fractions of Beta vulgaris

-Sample	Monovinyl Pchlide	Divinyl Pchlide
-		
	(%)	
Intact chloroplast	90±3	10±1
Thylakoids	91±2	9±0.5
Stroma	traces	traces
Outer envelope	87±4	13±2
Inner envelope	88±3	12±2

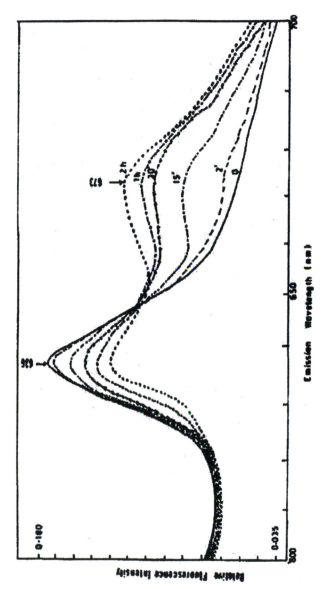

Figure 5. The room temperature fluorescence emission spectra of POR C catalyzed phototransformation of Pchlide to Chlide. Overexpressed POR C protein was incubated with Pchlide and NADPH for 15 min. It was illuminated with cool-white fluorescent light (20 $\mu mol\ m^{-2}\ s^{-1}$) for 0-120 minutes. The sample was excited at 433 nm and fluorescence spectra were monitored at room temperature (25^{0}C) in SLM-AMINCO 8000 spectrofluorometer. Spectra were not corrected for the photomultiplier tube response.

bacterial lysate in the absence of NADPH failed to phototransform Pchlide to Chlide in the presence of light (data not shown).

DISCUSSION

As compared to thylakoids, the relative pigment content of envelope membrane is quite low (Table II and III). Results demonstrate that Pchlide is mostly localized in the thylakoid and present in small amounts in the envelope. It is present only in trace amounts in the stroma. Had there been trafficking of Pchlide from the envelope to thylakoids or *vice versa*, significant amounts of Pchlide should have been present in the stroma. Therefore in the developed chloroplast, Pchlide present in the envelope may be synthesized *de novo* independent of thylakoids from protoporphyrinogen IX translocated from stroma. During greening Pchlide content (mg protein)$^{-1}$ decreased in thylakoids and increased in envelope membranes [37]. ALA biosynthetic enzymes are located in the stroma . Enzymes responsible for conversion of ALA to protoporphyrinogen IX i.e., ALA dehydratase, porphobilinogen deaminase, uroporphyrinogen decarboxylase and coproporphyrinogen oxidase are mostly located in the stromal phase [6,38]. Proto IX is present in the envelope and thylakoid membrane (Table-III). Therefore, protoporphyrinogen IX once synthesized in the stroma may associate either to the envelope or thylakoid membranes where they are oxidized to Proto IX. The presence of protoporphyrinogen oxidase is shown both in the envelope and thylakoid membranes [10,39]. Part of plastidic protoporphyrinogen IX migrates to mitochondria where it is oxidised by protoporphyrinogen oxidase II to Proto IX, the substrate of heme synthesis [40]. Substantial amount of Proto IX present in the stroma may be due to its own diffusion from the site of its synthesis i.e., envelope and thylakoids.

The next step in Chl biosynthesis is the conversion of Proto IX to Mg-proto IX by Mg-chelatase. The association of MP(E) with envelope membranes suggests that Mg-chelatase is functional in envelope membranes. It is often argued that chlorophyll biosynthetic enzymes present in envelope membranes are protein translocation intermediates. As there is an obligate requirement of three subunits of Mg-chelatase i.e., Chl D, Chl H and Chl I to assemble in a definite proportion to form the functional enzyme [41,42,43,44], it is unlikely that translocation intermediates of Mg-chelatase enzyme could mediate the synthesis of Mg-porphyrin. Chl H is localized in the stroma at low Mg^{2+} concentration (1 mM) and partially associates with envelope membranes at high Mg^{2+} concentrations (5 mM) and is clearly absent from thylakoids both in low as well as high Mg^{2+} concentrations [45]. As chloroplasts usually contain high salt which keeps their thylakoids stacked to form granum it is likely that Chl H associates with inner envelope membranes. Chl I is present in stroma and is not seen in membrane fractions [46]. A small part of Chl D comprising 110 amino acids is required for interaction with the partner subunits and maintenance of the enzyme activity *(44)*. Therefore, it is likely that assembly of Mg-chelatase may take place in the stroma in association with inner envelope membrane *(47)* Proto IX is converted to Mg-Proto IX in the stroma in close association with inner envelope membrane. This explains the presence of MP(E) in the envelope membrane and stroma.

It is likely that the amphiphilic tetrapyrrole Mg-Proto IX subsequently migrates to both envelope and thylakoids where it is independently esterified to MPE and subsequently metabolised to Pchlide. The uniform distribution of MV and DV forms of Pchlide suggests that there is a tight regulation of vinyl reductase enzyme in both envelope and thylakoid fractions. Both MV Pchlide and DV Pchlide are probably phototransformed to Chlide independently in envelope and thylakoid membranes. The presence of functional M g-Proto I X:s-adenosine m ethyl t ransferase and Pchlide oxido-reductase in the envelope membrane has already been demonstrated *(7,48-49)*. Further studies pertaining to the localization and functional enzymatic activity of envelope and thylakoid membranes with respect to MPE cyclase and vinyl reductase are needed to substantiate the proposed two different pathways of Mg-tetrapyrrole biosynthesis in envelope and thylakoid membranes. As stroma is hydrophilic in nature it is likely that stroma is not involved in the biosynthesis of the hydrophobic Mg-tetrapyrroles. Chl synthetase is localized in the thylakoids and absent from the envelope membranes *(50)*. Therefore, phytylation of Chlide to Chl takes place only in thylakoids and Chl is nearly absent from envelope membranes.

The mechanism of porphyrin synthesis in the outer envelope membranes is uncertain. As the amounts of pigments are invariably more in the inner than the outer, porphyrins are probably synthesized in the inner envelope membranes and subsequently migrate to the outer membrane. The extent of migration of pigments from the outer to the inner membrane may not be same for all pigments (Table-IV). As shown in Fig. 2, the ratio of Chlide/Pchlide (F678/F632) is also different in outer and inner envelope membranes. The mechanism of migration of pigments from inner envelope to the outer envelope membrane needs to be studied further.

Chloroplast may send signals to the nucleus and regulates nuclear gene expression *(51)*. The signal emanating from the chloroplasts could be plastidic Mg-Proto and MPE which are shown to regulate the nuclear gene expression of HSP70A *(52)*. Recently, GUN gene of the genome uncoupled *Arabidopsis* mutant which in contrast to wild type showed light-induction of nuclear LHCB gene expression in p resence o f t he c arotenoid b iosynthesis inhibitor, norflurazon is shown to encode for a Mg-chelatase H subunit *(53)*. In t he p resent study it is shown that signalling molecule(s) Proto IX, M g-Proto IX and/or its monoester are present in the outer envelope membrane. The sustrate of Proto IX i.e., protoporphyrinogen IX is shown to migrate out of chloroplasts *(54)*. Unlike other tetrapyrroles listed above, protoporphyrinogen IX lacks the rigid double bond conjugate system. Therefore, it is flexible and can swim all over the cell. This could diffuse to the cytoplasm and subsequently to the nucleus and regulate the expression of

126

nuclear gene involved in chloroplast biogenesis. Proto IX is found in the envelope membranes of mature green chloroplasts (55). Changes in the concentrations of Pchlide *(37)* and Proto IX and MP(E) (56) contents in the envelope membranes during plastid developmemt suggest a regulatory role of plant tetrapyrroles in chloroplast biogenesis. Diffusion of another tetrapyrrole i.e. protóporphyrinogen IX from chloroplast to mitochondria or plasma membrane has been shown *(40,54)*. Tetrapyrrole migrates from plastid to cytosol for phytochrome synthesis. Whether these tetrapyrroles diffuse independently or get associated with other proteins for their transport is an open question. Further studies need to be carried out to understand if these tetrapyrroles associate with other cytoplasmic and/or nuclear protein(s) for their transport and regulation of nuclear gene expression.

Results also demonstrate the presence of a different isoform of POR i.e. POR C. Different forms of POR are probably needed for Pchlide reduction for continued chlorophyll biosynthesis at different developmental stages and environmental conditions. The variability of *por* genes i. e. *por A, por B and por C* and their expression in dark and light in different species may have a bearing on the nature of seed germination i.e., hypogeal or epigeal and time the germinating seedling takes to emerge from sub-soil that receives very low light intensity to soil surface that gets full sun light.

ACKNOWLEDGMENTS

We wish to thank Professor Kenneth Keegstra, East Lansing, USA for OM 14 antibodies and Dr. A Matto, Beltsville, USA for LHCPII antibodies. Supported by a grant from University Grants Commission (F. 3-176/2001 SR-II).

REFERENCES

1. Eichacker, L. A.; Soll, J.; Lauterbach, P.; Rudiger, W.; Klein, R. R.; Mullet, J. E. *J. Biol. Chem.* **1990**, 265, 13566-13571.
2. Jilani, A.; Kar, S.; Bose, S.; Tripathy, B. C. *Physiol. Plant.* **1996**, 96, 139-145.
3. Kropat, J.; Oster, U.; Rüdiger, W.; Beck, C. F. *Proc. Natl. Acad. Sci. USA* **1997**, 94, 14168-14172.
4. Hoober, J. K.; Eggink, L. L. *Photosynth. Res.* **1999**, 61, 197-215.
5. Pineau, B.; Dubertret, G.; Joyard, J.; Douce, R. *J. Biol. Chem.* **1986**, 261, 9210-9215.
6. Manohara, M. S.; Tripathy, B. C. Planta **2000**, 212, 52-59.

7. Joyard, J.; Block, M.; Pineau, B.; Albrieux, C.; Douce, R. *J. Biol. Chem.* **1990,** 265, 21820-21827.

8. Walker, C. J.; Weinstein, J. D. *Proc. Natl. Acad. Sci. USA* **1991,** 88, 5789-5793.

9. Lee, H. J.; Ball, M. D.; Parham, R.; Rebeiz, C. A. *Plant physiol.* **1992,** 99, 1134-1140.

10. Matringe, M.; Camadro, J. M.; Block, M. A.; Joyard, J.; Scalla, R.; Labbe, P.; Douce, R. *J. Biol. Chem.* **1992,** 267, 4646-4651.

11. Holtorf, H.; Reinbothe, S.; Reinbothe, C.; Bereza, B.; Apel, K. *Proc. Natl. Acad. Sci. USA ,* **1995,** 92, 3254-3258.

12. Armstrong, G.A.; Runge, S.; Frick, G.; Sperling, U.; Apel, K. *Plant Physiol* **1995.** 108, 1505-1517.

13. Reinbothe, S.;Reinbothe, C.; Lebedev, N.; Apel, K. *Plant Cell* **1996,** 8, 763-769.

14. Oosawa, N.; Masuda, T.; Awai, K.; Fusada, N.; Shimada, H.; Ohta, H.; Takamiya, K. *FEBS. Lett* **2000,** 474, 133-136.

15. Walter, G.; Shalygo, N. V. *J. Photochem. Photobiol.* **1997,** 40, 175-182.

16. Tewari, A. K.; Tripathy, B. C. *Planta* **1999,** 208, 431-437.

17. Tewari, A. K.; Tripathy, B. C. *Plant Physiol.* **1998,** 117, 851-858

18. Keegstra, K.; Yousuf, A. E. *Methods Enzymol.* **1986,** 118, 316-325

19. Koike, H.; Yoshio, M.; Kashino, Y.; Satoh, K. *Plant Cell Physiol.* **1998,** 39, 526-532

20. King, T. E. *Methods Enzymol.* **1967,** 10, 216-225.

21. Simcox, P. D.; Reid, E. E.; Canvin, D. T.; Dennis, D. T. Plant Physiol **1977,** 59, 1128-1132

22. Coves, J.; Block, M. A.; Joyard, J.; Douce, R. *FEBS Lett.* **1986,** 208, 401-406.

23. Hukmani, P.; Tripathy, B. C. *Plant Physiol.* **1994,** 105, 1295-1300.

24. Rebeiz, C. A.; Matheis, J. R.; Smith, B. B.; Rebeiz, C. C.; Dayton, D. F. *Arch. Biochem. Biophys.* **1975,** 166, 446-465.

25. Hukmani, P.; Tripathy, B. C. *Anal. Biochem.* **1992,** 206, 125-130.

26. Tripathy, B. C.; Rebeiz, C. A. *Anal. Biochem.* **1985,** 149, 43-61.

27. Porra, R. J.; Thompson, W. A.; Kriedemann, P. E. *Biochim. Biophys. Acta* **1989,** 975, 384-394

28. Lowry, O. H.; Rosebrough, N. J.; Farr, A. L.; Randall, R. J. *J. Biol. Chem.* **1951,** 193, 265-275

29. Pattanayak, G. K.; Tripathy B. C. *Biochem. Biophys. Res. Commun* **2002,** 291, 921-924.

30. Li, H. M.; Moore, T.; Keegstra, K. *Plant Cell* **1991,** 3, 709-717.

31. Rebeiz, C. A; Montazer-Zouhoor, A.; Hopen. H. J., Wu, S. M. *Enzyme Microbiol. Technol.* **1984,** 6, 390-401.

32. Johanningmeier, U.; Howell, S. H. *J. Biol. Chem.* **1984,** 259, 13541-13549.

33. Chakraborty, N.; Tripathy, B. C. *Plant Physiol.* **1992,** 98, 7-11.

34. Tripathy, B. C.; Rebeiz, C. A. *J. Biol. Chem.* **1986,** 261, 13556-13564.
35. Tripathy, B. C.; Rebeiz, C. A. *Plant Physiol* **1988,** 87, 89-94.
36. Rebeiz, C. A.; Parham, R.; Fasoula, D. A.; Ioannides, M. In The biosynthesis of the tetrapyrrole pigments; Chadwick D. J.; Ackrill, K. Eds, Ciba Foundation Symposium 180; John Wiley and Sons, England, **1994;** pp. 177-193.
37. Barthelemy, X.; Bouvier, G.; Radunz, A.; Docquier, S.; Schimd, G. H., Franck, F. *Photosynth Res* **2000,** 64, 63-76.
38. Smith, B. B.; Rebeiz, C. A. *Plant Physiol.* **1979,** 63, 227-231.
39. Che, F. S.; Watanabe, N.; Iwano, M.; Inokuchi, H.; Takayama, S.; Yoshida, S.; Isogai, A. *Plant Physiol.* **2000,** 124, 59-70.
40. Lermontova, I.; Kruse, E.; Mock, H. P.; Grimm, B. *Proc. Natl. Acad. Sci. USA* **1997,** 94, 8895-8900.
41. von Wettstein, D.; Gough, S.; Kannangara, C. G. *Plant Cell* **1995,** 7, 1039-1057.
42. Papenbrock, J.; Gräfe, S.; Kruse, E.; Hänel, F.; Grimm, B. *Plant J.* **1997,** 12, 981-990.
43. Kannangara, C. G.; Vothknecht, U. C.; Hansson, M.; von Wettstein, D. *Mol. Gen. Genet.* **1997,** 254, 85-92
44. Grafe, S.; Saluz. H. P.; Grimm, B.; Hanel, F. *Proc Natl Acad Sci USA* **1999,** 96, 1941-46.
45. Nakayama, M.; Masuda, T.; Bando, T.; Yamagata, H.; Ohta, H.; Takamiya, K. *Plant Cell Physiol.* **1998,** 39, 275-284.
46. Nakayama, M.; Masuda, T.; Sato, N.; Yamagata, H.; Bowler, C.; Ohta, H.; Shioi, Y.; Takamiya, K. *Biochem. Biophys. Res. Commun.* **1995,** 215, 422-428.
47. Walker, C. J.; Willows, R. D. *Biochem J* **1997,** 327, 321-333
48. Joyard, J., Emeline, T., Miage, C., Berny-Seigneurin, D., Marechal, E., Block, M. A., Dorne, A., Rolland, N., Ajlani, G. Douce. R. *Plant Physiol.* **1998,** 118, 715-723.
49. Block, M. A.; Tewari, A. K.; Albrieux, C.; Marechal, E.; Joyard, J. *Eur J Biochem* **2002,** 269, 240-248.
50. Joyard, J.; Teyssier, E.; Miège, C.; Berny-Seigneurin, D.; Marèchal, E.; Block M. A.; Dorne, A. J.; Rolland, N.; Ajlani, G.; Douce, R. *Plant Physiol.* **1998,** 118, 715-723.
51. Oelmuller, R. *Photochem. Photobiol.* **1989,** 49, 229-239.
52. Kropat, J.; Oster, U.; Rüdiger, W.; Beck, C. F. *Plant J.* **2000,** 24, 523-31.
53. Mochizuki, N.; Brusslan, J. A.; Larkin, R.; Nagatani, A.; Chory, J. *Proc. Natl. Acad. Sci. USA* **2001,** 98, 2053-2058.
54. Gupta, I.; Tripathy, B. C. *Z. Naturforsch.* **1999,** 54C, 771-781.
55. Mohapatra, A. Tripathy, B. C. *Biochem Biophys Res Commun.* **2002,** 299, 751-754.
56. Mohapatra, A. Tripathy, B. C. *J. Plant Physiol.* **2003,** 160, 9-15.

Chapter 9

Green Remediation of Herbicides: Studies with Atrazine

Richard A. Larson, S. Indu Rupassara, and Seth D. Hothem

Department of Natural Resources and Environmental Sciences,
University of Illinois, Urbana, IL 61801

We report the use of higher plants to take up atrazine in hydroponic microcosms. The aquatic plant, hornwort (*Ceratophyllum demersum*), together with its associated microorganisms, was effective in removing atrazine . Breakdown products of atrazine in the presence of plants included deethylatrazine, deisopropylatrazine, and a glutathione conjugate of atrazine. Autoradiography indicated that only a fraction of labeled material remained within the plant cells, indicating that microorganisms (probably bacteria) played an important role in the disappearance of atrazine. Other higher plants, algae, and fungi capable of atrazine degradation were identified. The deethyl- and deisopropyl derivatives were also formed by photochemical processes in the absence of plants.

Phytoremediation, also called vegetation-enhanced bioremediation, is a potentially useful approach for long-term management of contaminated soils. Numerous studies in the literature report on the use of green plants to remove pesticides and other agrochemicals from soil (*1*). Plants may directly concentrate, metabolize, or volatilize potentially hazardous constituents. In addition, plants may foster aggregates of xenobiotic-degrading organisms in the vicinity of their roots (rhizosphere). Several studies in recent years have shown that some plants not only take up potentially hazardous organic materials, but metabolize them enzymatically to forms that may be less toxic, either to the plant itself or to the environment. It has been demonstrated that atrazine can be partly removed and converted to metabolites by the roots of hybrid poplar trees that could be planted in barrier strips adjacent to streams (2). Rice et al. (3) found that several herbicide-tolerant aquatic plants, namely *Ceratophyllum demersum*, *Elodea canadensis*, *and Lemna minor*, were able to reduce the concentrations of metolachlor and atrazine in test-tubes containing the plant and the herbicide.

Atrazine (structure 1) is a widely-used triazine herbicide that has been used by itself or in combinations for decades. Atrazine is recognized to have considerable environmental persistence, but at the same time its relatively high water solubility (35 mg/L, 1.6×10^{-4} M) has led to its detection in a variety of waters, including groundwater, surface freshwater, and drinking water (4-6). The highest levels in surface waters normally occur shortly after it is applied to fields in the spring. Early field studies demonstrated that atrazine was metabolized (or abiotically converted) to various products including deethylatrazine (structure 2), deisopropylatrazine (structure 3), and hydroxyatrazine (structure 4). These products continue to be the most frequently identified breakdown products of atrazine in the environment.

Losses of atrazine in water have been reported by several investigators. In one river (Roberts Creek, Iowa), atrazine disappeared rapidly in a light-dependent manner with half-lives ranging from about 1 day in summer to 7 days in winter (4). In contrast, the half-life of atrazine in a 230-m long wetland indoor mesocosm containing plants, *Daphnia*, frogs, and mosquito larvae ranged from 8-14 days (7).

Certain individual microorganisms or consortia of microorganisms are able to mineralize atrazine. Some are able to use it as a sole source of nitrogen. The partial metabolism of atrazine by particular microorganisms often results in the production of hydroxyatrazine (8), although other reports suggest that oxidative metabolites such as side-chain dealkylation products are also formed (9). A competing pathway in plants (and perhaps in some microorganisms) is a nucleophilic aromatic susbtitution reaction with glutathione (10) leading ultimately to a dechlorinated adduct (structure 5).

Studies of abiotic reactions of atrazine in water have focused on attempted remediation methods. Prados et al. (11) examined several different HO-radical-generating methods such as the Fenton reaction (iron[II]- hydrogen peroxide) and ozone-hydrogen peroxide combinations, and showed that triazines were removed with varying efficiencies by all such methods. Attack by HO• or ozone on atrazine is selective, leading principally to dealkylation products; addition to the triazine ring is a minor pathway (12-13). The N-deethylated product predominates over the deisopropyl compound by 4:1 in hydroxyl radical attack and by 19:1 in ozonation. The initial side-chain HO•-derived radicals, after diffusion-controlled addition of molecular O_2, rapidly eliminate HOO• to form Schiff bases. These rapidly hydrolyze to produce, in addition to the dealkylated amine, stoichiometrically equal quantities of acetaldehyde or acetone, respectively (*13*). Suggestions that abiotic (particularly photolytic) pathways for loss of atrazine in the aquatic environment may be important have been made (*4*). These authors pointed out a light-dependent relationship between atrazine degradation rate and day length. The rate constant was not significantly correlated with water temperature, but could have reflected biodegradation.

In an earlier study, we demonstrated that several North American aquatic plants showed promise in taking up atrazine from water without being killed in the process. Water primrose *(Ludwigia peploides)* and hornwort (*Ceratophyllum demersum*) were particularly efficient. Preliminary studies with *Ceratophyllum* indicated uptake of 50 μg/L atrazine with a half-life of 3-5 days (depending on the mass of plant material). Similarly, *Ludwigia* took up atrazine with a half-life of 6-7 days. *Ceratophyllum demersum*, also known as hornwort or coontail, is found in ponds and slowly flowing waters throughout North America (*14*). Typically, the species is found to inhabit deep pools and to predominate at >3 m depth (*15*). Preliminary work in our laboratory indicated that it was capable of removing dinitrotoluene and other nitrogenous organic compounds from water. It has a very large leaf area per gram of biomass, a characteristic similar to that of the subtropical aquatic species, parrotfeather (*Myriophyllum aquatica*), that has been studied for phytoremediation by Jeffers and Wolfe (*16*).

Results and Discussion

Atrazine and Green Plants

Loss of atrazine from containers having hornwort plants was faster (half-life typically 4-8 days) than that in the controls without plants, although the rate was greater in higher-light conditions. (Abiotic or photochemical loss of atrazine was also light-dependent, although slower.) Rate constants varied from 0.03/day in very low-light conditions to as much as 0.18/day at high light intensities (Figure 1). There was a slow degradation in the presence of plants under dark conditions. These results indicate that the presence of the hornwort plant enhanced the degradation process in a light-dependent manner.

Analysis of extracted plant material indicated that atrazine and one major radioactive metabolite was present in the plant. The metabolite had the same retention time (HPLC radiolabel analysis) and mass spectrum as the authentic

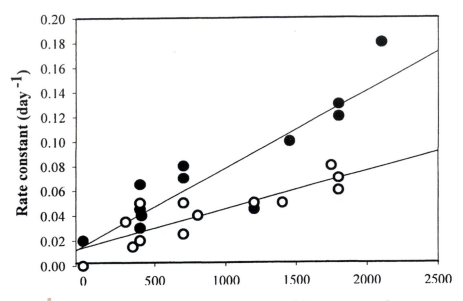

Figure 1. Atrazine loss under various conditions of illumination in the presence of hornwort plants (closed circles) and in the absence of plants (open circles).

atrazine-glutathione conjugate 5. LC-MS analysis indicated that a compound having a mass of 486, and fragment ions at 244 and 214, was present in the metabolite peak (these masses were not present in the control plant extract at the same retention time), with at least one other compound (having masses 478 and 316; these masses were present also in the control plant extract) co-eluting. The evidence further suggests that the metabolite is 5. Metabolism of atrazine in higher plants by conjugation with glutathione has been repeatedly demonstrated (17-18), and hornwort has shown glutathione-S-transferase activity as a mechanism of detoxification of xenobiotics (19).

In a parallel analysis of the growth medium, the major degradation product found was deethylatrazine (2). If the plant had been contributing to the formation of 2, it should have had a pool of the metabolite within the tissue, which was not observed. However, it is widely reported that 2 is produced by microbial degradation (9,20).

Autoradiographic analysis indicated that label was present inside the cells of the plant; label was more prevalent in the mature tissues, such as stems, than in the younger tissue. In quantitative terms, however, most radioactivity was found external to the plant, suggesting that surface-associated (epiphytic) microorganisms were responsible for the majority of the degradative activity.

Several grasses were screened. Annual (*Lolium multiflorum*) and perennial (*L. perenne*) ryegrass species were found to remove atrazine from solution with half-lives of 7-10 days. However, switchgrass (*Panicum virgatum*) and reed canarygrass (*Phalaris arundinacea*) were relatively ineffective; after 14 days less than 25% of the initially applied atrazine had been removed.

Degradation by Algae-Dominated Neuston

In some microcosms containing plants, a floating community dominated by algae formed. When this population was skimmed off and tested for its ability to degrade atrazine, a light-dependent degradation was observed with a half-life of 7-8 days. Although atrazine has been used to control algal blooms, the literature contains scattered reports of its degradation by resistant species (21).

Degradation by a *Penicillium* Fungus

A fungus found as a contaminant in an experimental attempt to sterilize reed canarygrass seeds was isolated and plated on agar. It was capable of degrading atrazine by about 80% within a week in the presence of the grass substrate;

however, if the grass was not present, it was less effective, with a half-life for atrazine loss of 12 days. Fungal metabolites included compounds 2, 3, possibly 4, and other (more polar) substances for which we had no reference standards. Microscopic examination of the fungus established that it was a member of the genus *Penicillium*. Degradation of atrazine by fungi has been reported previously (*22-23*).

Photodegradation (Abiotic)

Atrazine disappearance rates from abiotic "controls" having no plants were significant and light-dependent (Figure 1). In darkness, atrazine concentrations showed little or no decrease over a period of up to two weeks. Sunlight, greenhouse, or xenon lamp exposure, however, did result in significant losses of atrazine. Rate constants varied from about 0.03/day for low-light exposures (PAR = 300-400 µmol/m²/sec) to about 0.08/day for high-light periods (PAR = 1750). Since atrazine has no absorption of light in the solar UV or visible region, an indirect mechanism for photodegradation must be occurring. The principal sunlight-absorbing constituents of the hydroponic (Hoagland) medium are nitrate (Σ = 7.2 at 302 nm) and the micronutrient iron(III) chelate, Fe-DTPA (trade name Sequestrene-330; Σ = 7500 at its absorbance maximum, 278 nm; measurable extinction was observed out to *ca.* 450 nm).

Nitrate is well-known to produce hydroxyl radical on illumination (*24*), and indeed, experiments in which solutions of atrazine and nitrate were exposed to xenon-arc UV did lead to disappearance of atrazine (half-life ~30 hr). The iron complex also promoted degradation of atrazine upon illumination, although $FeCl_3$ did not. In previous studies, it was determined that in iron/organic compounds containing carboxylate groups, the iron species was easily reduced because of ligand-to-metal charge transfers (*25*). Such "photo-Fenton" reactions are known to produce reactive species such as hydroxyl radical.

To probe the role of hydroxyl radical in the abiotic photodegradation reaction, D-mannitol and *t*-butanol were added to the system. These compounds are widely used hydroxyl radical scavengers which have little activity toward most other oxidants (*26*). In separate tests, 0.8 mM of D-mannitol and 3 mM of *t*-butanol were added to 5% strength Hoagland's solutions containing 50 ppb of atrazine and illuminated. In both cases, the result was that there was almost no loss in atrazine for the duration of the test. This indicated that there were hydroxyl radicals present in the system. The two main products formed during the photodegradation reactions were deisopropylatrazine (3) and deethylatrazine (2). These compounds have been observed by numerous other investigators as the principal by-products formed when atrazine is treated with oxidizing agents.

Conclusions

Atrazine can be removed from hydroponic nutrient solution by a variety of organisms, including aquatic plants (hornwort), terrestrial plants (ryegrasses), bacteria associated with plants, algae, and at least one fungus (*Penicillium* sp.) Atrazine was also susceptible to photolysis in the presence of nitrate and complexed iron. Reaction products in these systems were dominated by side-chain oxidation products (deethyl- and deisopropylatrazine) and, in the case of hornwort, by a glutathione conjugate. A number of unidentified products, however, were also formed.

Methods and Materials

Plants

Plants were grown hydroponically in 5-10% Hoagland's nutrient solution (27). They were exposed to light under a variety of conditions; outdoor sunlight, xenon arc lamps filtered to approximate sunlight, growth chambers, and greenhouse. In all cases, UV and visible light intensities were monitored with Li-Cor (Lincoln, NE) meters. Temperature was also monitored, though it was shown in the case of hornwort that temperature had little effect on degradation rates.

Reagents and Analytical Methods

"Saturated" aqueous stock solutions of atrazine were prepared by stirring the solid with deionized-distilled water for 24 h and then filtering the undissolved material. The concentration of such solutions were determined by UV, comparing the absorbance to the molar extinction coefficient of atrazine at the wavelength of its maximum absorbance (223 nm). Atrazine concentrations in dilute solution experiments were measured using high-performance liquid chromatography (HPLC). A method used routinely in our laboratory, involving the injection of large volumes of solution on a crosslinked polystyrene column, allows us to measure concentrations as low as 5 μg/L. Analysis of the atrazine (and product) concentrations in each sample was completed using HPLC (Beckman 110B solvent delivery module) with an absorbance detector (Spectroflow 757) attached. Injection volumes were 2.0 mL. The column used was an Altima C18-3u with a mobile phase of 1 MeCN:1 H_2O and a flow rate of 2 mL/min. A more rapid technique was also developed, using a novel HPLC column (Rocket®, Alltech), which permitted us to analyze atrazine-containing samples in 6 min as opposed to 25 min for a conventional column; however, this method did not permit complete separation of reaction by-products.

LC-MS Analysis

The atrazine-glutathione conjugate (5) was tentatively identified using electrospray LC-MS-MS (Thermo Finnigan, San Jose, CA) and other techniques. An authentic standard of the conjugate (28) was synthesized and analyzed by HPLC-UV, proton NMR, and fast atom bombardment MS. The presence of the compound at low levels in hornwort tissues was essentially confirmed by extracting the pulverized plant material with several portions of methanol, followed by solid-phase extraction using Oasis HLB cartridges (Waters, Milford, MA), and LC-MS. During mass spectrometric analysis, an LC peak in the extract contained masses at m/z 487 (M+1), 244, and 214 and eluted at the same time as the authentic specimen. The extract, contained at least one other material with ions at m/z 479, 317, as well as others; however, the sole source of the characteristic fragment ions of the conjugate was its apparent molecular ion, m/z 487.

Autoradiography

Hornwort plants were kept in Hoagland's nutrient solution containing 1 mg/L UL-^{14}C-atrazine for three weeks in a growth chamber under day/night conditions. Control plants were kept in the same manner in the presence of cold atrazine. Samples for autoradiography were dehydrated, embedded in paraffin, and cut to 4-8 microns thickness using a rotary microtome. The sections were placed on microscope slides and dipped in emulsion and developer, then observed microscopically for the presence of black silver metal deposits.

Atrazine-degrading Fungus

The fungus was isolated and maintained on potato dextrose agar (PDA) plates. A macroscopic surface view of its colonies was deep green in color, with a white mycelial halo; the reverse (underside) of the colony was pale yellow. Fungal specimens were also examined on glass slides for microscopic observation of conidial and vegetative structures. To investigate its ability to degrade atrazine, loss of atrazine from liquid Hoagland's medium supplemented with 50 ppb of the herbicide was monitored after the addition of a 1-cm diameter fragment of the PDA medium containing the fungal mycelium. These samples were incubated in a growth chamber at 30±2° with day-night illumination.

Photodecomposition of Atrazine in Hoagland's Nutrient Solution

Each of the solutions to be tested was poured into 13 x 100 mm borosilicate test tubes. The test tubes were placed on a sample rotator under a solar simulator containing a 2500-watt xenon arc lamp (Optical Radiation Corp, Azusa, CA). The lamp used contained a filter to remove all wavelengths less than 290 nm. The distance from the lamp to the samples was adjusted to simulate midday solar intensity at 40° latitude. The solar simulator was turned on 20 minutes prior to the start of the experiment to allow it to come to a steady state. Light readings in the photosynthetically active region (PAR) and UV-B range were taken using Li-Cor probes. The average PAR reading for the atrazine tests was 1450 $\mu M/m^2$ sec, while the UV-B averaged 0.37 W/m^2. For each hour that the tests were run, one of the test tubes was removed from the rotator and stored in dark conditions until analysis. Samples that received no light exposure were also taken in order to provide initial concentrations. All of the tests lasted for 4 or 5 hours.

Acknowledgments

We thank Karen Marley, Diane Pedersen, Gerald Sims, Robert Twardock, and Liqiang Zhou for discussions and assistance, and the Illinois Hazardous Waste Center, Illinois Council on Food and Agricultural Research, and Illinois Groundwater Consortium for financial support.

References

1.   Kruger, E. L., T. A. Anderson, and J. R. Coats, eds.   1997. *Phytoremediation of soil and water contaminants*. ACS Symposium Series #664, American Chemical Society, Washington, DC.

2.   Burken, J. G., and J. L. Schnoor. 1997. Uptake and metabolism of atrazine by poplar trees. Environ. Sci. Technol. 31:1399-1406.

3.   Rice, P. J., T. A. Anderson, and J. R. Coats. 1997. Ch 10: Phytoremediation of herbicide-contaminated surface water with aquatic plants. pp. 133-151 *in* Kruger, E. L., T. A. Anderson, and J. R. Coats, eds., *Phytoremediation of Soil and Water Contaminants*, ACS Symposium Series #664,. American Chemical Society, Washington, DC.

138

4. Kolpin. D. W. and S. J. Kalkhoff. 1993. Atrazine degradation in a small stream in Iowa. Environ.Sci. Technol. 27:134-139.

5. Kolpin, D. W., S. J. Kalhoff, and D. A. Goolsby. 1997. Occurrence of selected herbicides and herbicide degradation products in Iowa's ground water, 1995. Ground Water 35: 679-688.

6. Battaglia, W. A. and D. A. Goolsby. 1999. Are shifts in herbicide use reflected in concentration changes in midwestern rivers? Environ. Sci. Technol. 33:2917-2925.

7. Detenbeck, N. E., R. Hermanutz, K. Allen, and M. C. Swift. 1996. Fate and effects of the herbicide atrazine in flow-through wetland mesocosms. Environ. Toxicol. Chem. 15:937-946.

8. Mandelbaum, R. T., D. L. Allan, and L. P. Wackett. 1995. Isolation and characterization of a *Pseudomonas* sp. that mineralizes the S-triazine atrazine. Appl. Environ. Microbiol. 61:1451-1457.

9. Behki, R. M. and S. U. Khan, S. 1986. Degradation of atrazine by *Pseudomonas*: N-dealkylation and dehalogenation of atrazine and its metabolites. J. Agric. Food Chem. 34:746-749.

10. Field, J. A. and E. M. Thurman. 1996. Glutathione conjugation and contaminant transformation. Environ. Sci. Technol. 30:1413-1418.

11. Prados, M., H. Paillard, and P. Roche. 1995. Hydroxyl radical oxidation processes for the removal of triazine from natural water. Ozone Sci. Eng. 17:183-194.

12. Acero, J. L., K. Stemmler, and U. Von Gunten. 2000. Degradation kinetics of atrazine and its degradation products with ozone and OH radicals: a predictive tool for drinking water treatment. Environ. Sci. Technol.. 34:591-597.

13. Tauber, A., and C. von Sonntag. 2000. Products and kinetics of the OH-radical-induced dealkylation of atrazine. Acta Hydrochim. Hydrobiol. 28:15-23.

14. Winterringer, G. S. and A. C. Lopinot. 1977. *Aquatic plants of Illinois.* Illinois State Museum, Springfield, IL.

15. Best, P. H. 1977. Seasonal changes in mineral and organic components of *Ceratophyllum demersum* and *Elodea canadensis.* Aquat. Bot. 3: 337-348.

16. Jeffers, P. M. and N. L. Wolfe. 1997. Degradation of methyl bromide by green plants. In *Fumigants: environmental fate, exposure, and analysis.* J. N. Seiber, J. A. Knuteson, J. E. Woodrow, N. L. Wolfe, M. V. Yates, and S. R. Yates, eds. ACS Symposium #652, pp. 53-59. Amer. Chem. Soc., Washington, DC.

17. Gimenez, E. R., E. Romera, M. Tena, and R. D. Prado 1996. Fate of atrazine in treated and pristine accessions of three *Setaria* species. Pestic. Biochem. Physiol. 56:196-207.

18. Hatton, P. J., D. Dixon, D. J. Cole, and R. Edwards. 1996. Glutathione transferase activities and herbicide selectivity in maize and associated weed species. Pestic. Sci. 46:267-275.

19. Pflugmacher, S., and C. Steinberg. 1997. Activity of phase I and phase II detoxication enzymes in aquatic macrophytes. Ver. Angew. Bot. 71:144-146.

20. Roberts, T. R. 1998. Part 1: Herbicides and plant growth regulators. *In Metabolic pathways of agrochemicals* (T. R. Roberts, D. H. Hutson, P. W. Lee, P. H. Nicholls and J. R. Plimmer, eds.), pp. 629-635. The Royal Society of Chemistry, Cambridge, UK.

21. Zablotowicz, R. M., K. K. Schrader, and M. A. Locke. 1998. Algal transformation of fluometuron and atrazine by N-dealkylation. J. Environ. Sci. Health, B33:511-528.

22. Kaufman, D. D. and J. Blake. 1970. Degradation of atrazine by soil fungi. Soil Biol. Biochem. 2:73-80.

23. Donnelly, P. K., J. A. Entry, and D. L. Crawford. 1993. Degradation of atrazine and 2,4-dichlorophenoxyacetic acid by mycorrhizal fungi at three nitrogen concentrations in vitro. Appl. Environ. Microbiol. 59:2642-2647.

24. Zepp, R. G., J. Hoigné, and H. Bader. 1987. Nitrate-induced photooxidation of trace organic chemicals in water. Environ. Sci. Technol. 21:443-450.

25. Miles, C. J., and P. L. Brezonik. 1981. Oxygen consumption in humic-colored waters by a photochemical ferrous-ferric catalytic cycle. Environ. Sci. Technol. 15:1089-95.

26. Buxton, J. V., C. L. Greenstock, W. P. Helman, and A. B. Ross. 1988. Critical review of rate constants for reactions of hydroxyl radicals in aqueous solution. J.. Phys.Chem. Ref. Data 17:513-886.

27. Hoagland, D.R., and D.I. Arnon. 1938. The water-culture method for growing plants without soil. Calif. Agr. Ext. Public. 347.

28. Frassanito,R., M. Rossi, L. K. Dragani, C. Tallarico, A. Longo, and D. Rotillo. 1998. New and simple method for the analysis of the glutathione adduct of atrazine. J. Chromatogr. A795:53-60.

Chapter 10

Enzymatic Microbial Degradation: In-Process Bioremediation of Organic Waste-Containing Aqueous Solvents

William M. Nelson[1], Valerie Tkachenko[1], Tim Delawder[2], and Dan Marsch[1]

[1]Waste Management and Research Center, 1 Hazelwood Drive, Champaign, IL 61820–7465
[2]SWD, Inc., 910 South Stiles Drive, Addison, IL 60101–4913

Chemical waste streams are an environmental problem facing all chemical industries. It is shown that using live microbes to remediate oil contaminated aqueous waste streams is an example of green chemistry. Commercially-available microbes have been successfully used in military garages and metal finishing companies to continuously remediate cleaning solutions. The potential application of this technology to petrochemical plants, chemical plants, refineries, food processing plants, marine barges, truck washes, wood treating plants and ground water remediation applications are explored. In our work we have examined the effectiveness of microbes to remediate cleaning solutions. The microbes can remediate crude oil, oils, solvents, greases, amines, cresol, phenols, and others hydrocarbon compounds. The systems and our results are described.

Introduction

Green solvents, like water, are more environmentally benign.[1] Due to improper disposal and accidental spills, aqueous solutions containing non-halogenated solvents (including alcohols), oils and PAHs have become environmental contaminants. These chemicals are toxic and many are suspected carcinogens.[2] Substantial research has been conducted to find solutions to these contamination problems. Many of the remediation technologies that have been developed have limitations including long clean-up times, difficulty stimulating and controlling the *in situ* growth of the degrading microbes, and high construction and operating costs. The principle tenet of green chemistry, pollution prevention, advises that preventing the pollution is preferable to managing it after it is produced.[3]

Aliphatic compounds, notably the greases, oils, etc. are common contaminants in soil and groundwater and inadvertently or casually released into the environment, notably to the soil and groundwater. When released, the compounds physically and chemically interact, forming non-aqueous phases, flowing downward, sorbing strongly with soil organics and minerals and dissolving into groundwater. While these aliphatic compounds are often recalcitrant, under favorable conditions they can be transformed and degraded through microbial-mediated processes. Naturally occurring microorganisms have been shown to degrade chlorinated and non-halogenated solvents, oils, and even PAHs.[4] The latter are relatively persistent and are more difficult to be degraded than many other organic contaminants under natural conditions.[5] Nevertheless, in the environment the effectiveness of the biological degradation of these solvents is often limited by microbial competition, aquifer temperature, the availability of nutrients, and other factors impacting the growth and activity of microbes.[6] We are looking at how an agricultural process, enzymatic microbial degradation, might be advantageously used as a green chemistry tool to reduce pollution resulting from aqueous solvents containing hydrocarbon contaminants.

Biodegradation of hydrocarbons

In the metal finishing industry and in other industries requiring the removal of contaminants from metal surfaces, there have been strong environmental reasons to move to alternative solvents, especially aqueous technologies.[7] Often the aqueous systems have been harsh, combining extremes of pH, high heat and physical cleaning.[8] However, under suitable conditions, a milder technology can be equally employed. Furthermore, these systems can be "self-remediating."[9]

Commercial bioremediated cleaning systems typically employ a mildly alkaline cleaning solution and control system that utilizes microbes in the solution to consume the oil/grease that is removed from metal surfaces during the cleaning process. The system operates at relatively low temperatures (104°F – 131°F) (40°C - 55°C) and a pH range of 8.8 - 9.2, which is a viable habitat for these microorganisms.[10] The test results and a review of historical operating records at many companies (like SWD) show that the microbe-remediation systems provide an environmental benefit by eliminating the need for alkaline bath disposal, thereby extending the bath life and reducing the amount of liquid and solid wastes produced by the cleaning operation.[11] The economic benefit associated with this technology is low operating and maintenance labor and reduced chemical costs, and the payback period can be less than a year.

This technology, however, is not a panacea and its health effects and broad cleaning efficacy must be rigorously evaluated.

Cleaning Technology

The cleaning process actually takes place in two separate operations. When parts come in contact with the solution, the oil and impurities are emulsified into micro-particulates. The particulates are then consumed by microorganisms, which are present in the bath or spray. The microbe consumption of the oil present in the bath, as its food source, results in the production of CO_2 as a by-product. The contaminant organics <u>must</u> be feedstocks for the microbes or this remediation will not work.

Bioaugmentation is the application of biological treatment for cleanup of industrial wastes and hazardous chemicals.[12] is accomplished by adding specific microbes to enhance the degradation of toxic substances. Bioaugmentation is particularly useful for degrading oil or grease. It is growing in popularity for treating drains and grease traps at commercial kitchens and food processing plants, and now in eliminating the contaminants removed from surfaces by cleaning technologies.

Hydrocarbon-eating microbes are blended with special nutrients and catalysts and then introduced into petroleum-contaminated water. The microbes break down the hydocarbons that the surfactants sequester.

Chemistry

The idea of using microbes to consume oil is not revolutionary: For over 40 years microbes have been utilized to consume oil from oil spills. The novelty in our case results from its use "in-process" as a green chemistry application. The microbial remediated system requires an effective cleaner in tandem with an effective remediation technology. Most conventional alkaline cleaning solutions would immediately kill the oil-consuming microbes, because of high operating temperatures or high pH. The cleaning chemistry was constructed around the characteristics of the microbe.

Cleaning solvents like mineral spirits dissolve contaminants, removing them from a metal surface. A product containing remediation microbes attempts to sequentially accomplish two tasks: (1) to remove contaminants from the surface (via surfactants that are not feedstocks for the microbes); and (2) to biodegrade (remediate) the contaminants, thereby preventing their buildup.

The first step involves finding an effective cleaning chemistry that removes the oil and grease. It might also include mechanical cleaning (for example, use of a parts brush or adjustable faucet) to aid in the removal of grease and oil.

Microbiology

In the second step, the microbes in the solution then break down the hydrocarbons (grease, oil, etc.) into low chain length hydrocarbons, and ultimately water and carbon dioxide (CO_2). The CO_2 is released into the air, and the water is eliminated through evaporation.[13]

The chemical solutions include a surfactant, nutrients, and buffer solutions. The cleaner is used to break the bond between the part and the oil and then form a micelle around the oil particle. The surfactant aids the cleaning process. One buffer contains phosphoric acid and nutrients for the microbes. The other buffer contains sodium hydroxide and nutrients for the microbes. The buffering solutions are used to maintain the cleaning solution pH, as well as supply nutrients for the microbes. The microbes ingest the oil first, but if the oil concentration in the cleaning solution is low, the microbes eat what is available. To prevent the microbes from eating the cleaner or nutrients are added in the buffer solutions as a supplementary food source.

The microbes attach to the oil droplet or micelles and begin secreting discharging enzymes that break down the hydocarbon structures into more water soluble, digestible materials that are subsequently absorbed through the cell wall and digested further. Nutrientss mixed with the microbes accelerate up the organisms' rate of reproduction and digestion. When provided a supportive environment, the microbes reproduce and disperse throughout the contaminated water, increasing the overall biomass of microbes in an exponential manner until all of the available hydrocarbons are consumed.

As with all biological systems, there is some time lag for microbes to grow in response to the introduction of nutrients. This technology will need to be adjusted for individual applications.

Microbes currently used in commercial systems are completely safe to humans and the environment. The strains of microbes present in the systems are all classified as American Type Culture Collection (ATCC) Class I. Organisms in this classification have no recognized hazard potentional under ordinary conditions of handling. They are subject to unrestricted distribution by the ATCC, U.S. Department of Health, Public Health Service and the Toxic Substances Control Act (TOSCA). Within a plant or work environment, the microbes do not pose a health threat, but the bath conditions can potentially support the growth of adventitious microorganisms (Osprey Biotech, personal communication.)

Engineering

Within the cleaning system, the solution temperature, pH level, and additions of biodegradable compounds are controlled. The cleaning solution is circulated continually between the cleaning tank and the processing tank. In the ChemFree system the solution is self-contained (in other words, it contains all nutrients needed to support microbe growth.) In the SWD system, a processing tank's automated control system constantly monitors the bath solution and maintains a preset concentration by adding supplemental chemical solutions as needed.

We used two different cleaning geometries: a parts washing "sink" (about 10 gallons, ChemFree, Fig. 1) and a larger cleaning bath (200 gallons, Mineral Masters, Fig. 2).

Figure 1. ChemFree SmartWasher.

Figure 2. BioClean Separator Module I[11]

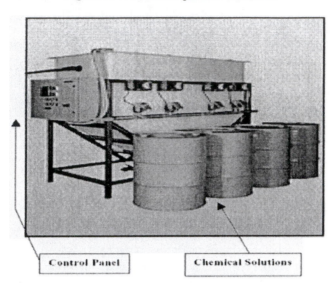

Control Panel   Chemical Solutions

Results and discussion

Facing increasing restrictions on the use of traditional organic solvents, the cleaning industry has looked carefully at aqueous cleaning technologies as a potential alternative. The limitations of aqueous cleaning have emerged, and among them is contaminant buildup (which reduces performance.) Depending upon the complexity of the contaminants, bioremediation can reduce their accumulation by breaking them down. Examples of microbial cocktails bioremediating polyaromatic contaminants[14] are indications that there is substantial potential for this technology in cleaning. The metabolic breakdown by microbes of both aliphatic and aromatic hydrocarbons has a common fatty-acid-formation stage. During this process the oxidation requires specific enzymes, which are produced by a few microorganisms. The key, obviously, is that these microbes are specifically targeted for the contaminants.

While the microbes are invisible to the naked eye, they must be packaged in such a way that they are effectively released.[15] Being living organisms, the microbes will grow in number as long as the contaminant ("food") and nutrients are present. During their lifetime these organisms require substrate, minerals and trace elements to remain viable. Other factors that affect the health of these microorganisms include:

- Hydraulic load (amount of water entering the system);

- Contaminant concentration;
- Presence of poisons or harmful chemicals (to microbes);
- Oxygen (presence or absence is important);
- Temperature (tepid, around 100 °C);
- pH (usually 6.5 – 8.5).

Smart Washer (ChemFree)

The Smart Washer sinks were utilized in garages by the Illinois National Guard. Routinely the military personnel cleaned disassembled motor vehicles during maintenance repairs. The work load varied: the sink might be used continuously for several days, then used sporadically the next few days. The levels of non-remediated hydrocarbons is measured by FOG (Fats, Oils and Grease) measured in g/mL. High FOG levels are indicative of non-remediated oils, high levels of recent cleaning activity or contaminants that are not substrates for the microbes.

The ChemFree units were located in 5 geographically different locations in Illinois (called "sites"). The work at each site was vehicle maintenance.

Data summary

Results from dipslides were used to estimate colony density, the number of colonies present per ml of cleaning solution called Ozziejuice® (measured in cfu). Table 1 shows the results for each SmartWasher over an eight-week period. These values were determined weekly, and solely indicate the cfu/mL at the time the sample was taken.

Table 2 shows the results from the Analytical Chemistry lab at the Waste Management and Research Center (WMRC) for the concentrations of oil and grease present in the Ozziejuice. The concentrations are reported in grams per liter.

Two additional sets of data were collected (although not reported here), that relate to the use and performance of the SmartWasher system. The perceived level of cleaning and the temperature were both recorded each week. It was determined that neither the level of activity nor the temperature on a specific day was as important as maintaining the recommended SmartWasher conditions during the entire week.

Discussion

One of the initial assumptions when the study began was the possibility of validating the existence of a correlation between the cfu levels and levels of fats,

Table 1. Number of Colony forming units in the Ozziejuice® samples (per ml) [cfu/mL]

Site	Week 1	Week 2	Week 3	Week 4	Week 5	Week 6	Week 7	Week 8
I	0	0	0	0	10 E4	10 E3	10 E6	10 E5
II	10 E3	10 E4	Na	10 E5	10 E5	10 E5	na	10 E4
III	10 E5	10 E4	10 E5	10 E4	10 E4	10 E4	10 E4	10 E4
IVa	10 E5	10 E5	10 E6	10 E6	10 E6	10 E6	10 E6	10 E6
IVb	10 E7	10 E5	10 E6	10 E6	10 E6	10 E6	10 E6	10 E6
IVc	na	10 E7	10 E7	10 E7	10 E7	10 E6	10 E6	10 E7
V	10 E7	10 E6	10 E6	10 E6	na	10 E6	10 E7	10 E7

- - - - -   Indicates when the microbes were added.

na   Data not collected that week

Table 2 Oil and Grease Levels in Ozziejuice® (g/L)

Site	Week 1	Week 2	Week 3	Week 4	Week 5	Week 6	Week 7	Week 8
I	45	>130	>98	76	>91	>110	>110	>96
II	52	46	Na	8.7	19	19	na	>47
III	33	38	31	29	62	17	16	22
IVa	7.5	5.4	3.7	2.4	1.7	2.8	3.4	3.2
IVb	0.97	0.95	1.0	1.1	1.1	1.0	1.1	1.4
IVc	na	0.86	1.7	1.4	0.99	1.0	0.87	0.99
V	9.3	1.5	2.7	2.9	na	5.7	4.7	3.6

oils and grease (FOG). A continuous reading of the week's values would be helpful to determine this.

From the data it is clear that all but one site were able to either maintain a low level of FOG or reduce the level from initial values. This was accomplished while operating on a regular cleaning schedule, including irregular addition of hydrocarbon contaminants. The SmartWashers had small valume changes due to evaporation and "drag-out." The volume loses were compensated only by the addition of more Ozziejuice.

Two factors, high levels of microbe population and low concentrations of contaminants, are postulated to affect good cleaning performance. Sites IVa,b,c and V maintained high cfu counts (>10E5) and these correlated with low FOG levels (<9 g/L). Visual examination of the solutions at these sites revealed that they began as amber and became dark-brown in color, but clear and not opaque. It was observed that these same sites had recorded high satisfaction levels on the part of the operators (determined from operators satisfaction with cleaning performance and the mildness of the solutions compared with previously used organic solvents), and they seemed to maintain the SmartWasher properly(detailed maintenance logs).

The reciprocal scenario of lower cfu (<10E4) and higher FOG (>20 g/L) was also observed. These sites also had either operator uncertainty or dissatisfaction. It must be noted that the users were never given the FOG or cfu values during the testing period, so this did not influence the sites' performances. These levels were seen most prominently at the "Site I" and to a smaller degree at both "Site II" and "Site III". These latter sites (I-III) did not report having heavier activity than the other sites in terms of parts washing.

Interestingly, it was found that several sites had Several SmartWashers that never degraded the contaminants to very low concentrations (<10 g/L) during the test period. Since the study with the sites did not include starting with new SmartWashers, we cannot speculate on whether the level would return to zero in these units. In a further study, this would be a valid point to observe. ChemFree maintains that complete remediation of the cleaning solution would take about seven days.[9] We speculate that there may also be a contaminant that entered the system causing the microbial level to remain low. The contaminant may have not been degraded by the bacteria population or possibly may require more time.

Based upon the responses on the weekly questionnaire, the level of satisfaction on the part of the users increased at all sites. This observation is germane to the successful adoption of any new technology: increasing familiarity and comfort with a new technology will increase the likelihood of users continuing to use the technology. There are several possible reasons for this increase, including increased comfort with a new technology and "hands-on" results. There is still some room for improvement with the use of the ChemFree system at all sites.

Following up on the last point, in order to increase the comfort level on the part of the workers, a means of assaying the level of microbes would be helpful.

As the ChemFree system is marketed (and most others as well), this is not possible. The system works like a "black box". Throughout the study, we allowed the users of the SmartWashers to add microbes on their own schedule (as the microbes either die, do not adequately repopulate or become weak over time). We found that when the SmartWasher is working effectively, the levels of bacteria did not decrease after 4 weeks, but we did not measure the "health" of the microbes. A close relationship between the FOG and microbe population levels was not universally valid in this study. In general, we did observe that the oil and grease concentrations were low in most cases when the bacteria level was at or above 10 E6 colonies per ml. As has been reported[9] the population and lifetime of the microbes in any one washer will depend upon several factors, and these are closely related to use and proper maintenance of the system. Establishing a test that could measure the "health" of the "microbial remediation"-based cleaning system would be beneficial.

SWD (Mineral Masters)

SWD is a metal-finishing company in Chicago, IL. As part of their production process, they clean metal parts in barrels prior to plating. The company installed a cleaning system utilizing the Mineral Masters bioremediating cleaning technology.

Though larger in volume, the technology still involves contaminant isolation followed by remediation. Unlike the ChemFree system, however, the microbes and nutrients must be vigilantly monitored to insure that remediation occurs. As was shown in the diagram (Fig.2), this system is more open and relies upon operator attention to maintain its proper cleaning efficacy.

Discussion

The data presented in Table 3 represent two separate 5 month testing periods (Series 1 and 2). The data show that the FOG level begins to rise almost immediately and appears to "leve" between 12-14 g/mL. While this is only slightly higher than the optimum values in the ChemFree system, the characteristics of the SWD solution at this point are different. Though the value is only 12-14, the solution is dark and "mud-like" in appearance and has a strong smell of lubrication oils. The data ends after 5 months, as the cleaning bath is no longer useful. This data set reveals several relevant details regarding the use of microbial baths.

The work done at SWD is not as "controllable" as the work in the SmartWashers. This means that the identity of the contaminants is not always known. This places the requirement on the system that there be a sufficient

Table 3. Fats, oils and grease measurements taken over 5 months use of a cleaning solution

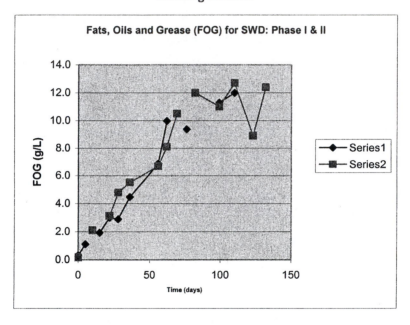

variety of microbial strains to "feed" on the contaminants. The fact that the FOG level continued to rise, even though the cfu/mL was maintained at 1E6 (data not shown) would mean that the contaminants were not being remediated by the microbes in solution. Changes made between Series 1 data and Series 2 data had no effect on the rate of the bath's deterioration.

The differences between Series 1 and 2 included using filtered water, increasing the dissolved oxygen in solution and periodically removing sludge build-up. None of these worked to slow the initial rapid deterioration of the bath's performance.

The work at SWD involves two shifts and it was difficult to determine the frequency of the bath's use and the nature of the parts cleaned in it. Under test conditions at WMRC the Mineral Masters cleaning solution (called Scumbugs®) remediated a solution with a known set of contaminant hydrocarbons.

Outlook

Although nutrient concentration plays a vital part in microbial cultivation, an increase in solvent concentration and an excessive range of contaminants causes stress within the system, indicated by a longer lag phase and poorer remediation. Cytotoxic effects may occur at higher solvent concentrations, which may explain the decrease in the calculated specific growth rate following a doubling of contaminant concentration. Beyond the cleaning sink and bath, biodegradation studies using bacterial strains have been carried out on a wide range of solvent and hydrocarbon pollutants, such a bisphenols, gasoline oxygenates, phenol, trifluoromethane and dimethylsilanediol.[16] The positive results reinforce the potential application to many industrially relevant bioremediative technologies.

Beyond industrial cleaning

2-propanol, or isopropyl alcohol (IPA), production worldwide exceeds 10^6 tons per year. It is used industrially as a solvent, intermediate, de-icer and has many other applications since it is cheaper than ethanol.[17] Significant amounts of toxic waste streams are also generated in the production of rubber, cosmetics, textiles, pharmaceuticals and fine chemicals. Difficulties in the complete detoxification of pharmaceutical waste streams, for example, occur due to the presence of a wide range of organic solvents and xenobiotic compounds, which may not be metabolized by microorganisms in conventional treatment plants. It has been proven that compounds containing isopropyl groups are moderately resistant to microbial breakdown, suggesting that IPA-containing waste streams may be more problematic than propanol-containing streams within the ecosystem.[18 ,19]

Chlorinated Solvent Remediation

An important class of transformations occurs in anaerobic environments. Many of the transformations are reductive, and many yield useful energy to specific anaerobic bacteria. They include reductive dechlorination, dehydrochlorination and dichloroelimination. Of these, reductive dechlorination is often a growth-supporting reaction, while the others may be abiological or catalyzed by biological molecules. The reactions may result in chlorinated products, but there are often reaction sequences leading to completely dechlorinated products. The behavior of carbon tetrachloride (CT), 1,1,2,2-tetrachloroethane (TCA) and the chloroethenes, perchloroethylene (PCE) and trichloroethylene (TCE), illustrate the range of anaerobic transformations that are possible, as well as the limited transformation that often is seen in the environment.

A laboratory study, using anaerobic sludge that had been fed chlorinated compounds, a cell-free extract from the sludge, and killed controls, showed that TCA was transformed to four products and that these were further transformed, suggesting that it might be possible to degrade TCA to innocuous products. Reductive dechlorination of PCE and TCE has been studied in many laboratories.[20] Selected field studies of anaerobic transformations help delineate the applications of this type of bioremediation in "in-process" green chemistry.

Anaerobic metabolism

The anaerobic transformations, by and large, are reductions of the contaminant compound and substitutions for chlorine. Most reactions fit into four categories: the first two are substitutions, but not reductions; the latter two are also reductions, each involving two electrons.

Isolated organisms that are capable of reductive dechlorination of chlorinated compounds are known. It is clear under which conditions that optimal dechlorination occurs and when incomplete dechlorination occurs due to suboptimal physiochemical conditions, deficiencies in electron donors or nutrients present, or a lack of appropriate organisms.[21]

Aerobic metabolism

Aerobic cometabolism using methanotrophic bacteria has shown promise for remediation of chlorinated solvent-contaminated groundwater that may pose a threat to human health and the environment. Cometabolism involves the gratuitous degradation of a nongrowth substrate, such as a chlorinated solvent, because of the presence of a nonspecific oxidizing enzyme. The cometabolizing bacterium used in this study was the mutant methanotroph *Methylosinus trichosporium*.[22]

PAH Biodegradation

Bioremediation of coal tar-contaminated soils containing polycyclic aromatic hydrocarbons (PAHs) is highly challenging because of the low solubility and strong sorption properties of PAHs. The biodegradation of various PAHs have been assessed.

PAHs are relatively persistent and recalcitrant in soils and are more difficult to be degraded than many other organic contaminants under natural conditions,[5] In addition, since PAHs are hydrophobic compounds with low solubility in water, they have a greater tendency to bind with organic matter or soil, limiting their availability to microorganisms. Additionally, PAHs may diffuse into the micropores of the soil, making it unavailable for biodegradation.[23] To enhance biodegradation, researchers have used surfactants [24] and organic solvents[25] to improve the availability of PAHs.

Conclusions

For the hydrocarbon compounds discussed in this paper, it has been shown that transformation to innocuous products is possible, but the proper microbes and the proper conditions (including operator acceptance) must be used. There is abundant evidence that complete transformations are not usually seen, but improvements (i.e. slowing of the deterioration of the solution) can be achieved.. Complete conversion, when it is found, probably depends critically on the key components:

- Proper physical conditions (temperature, chemicals, air); and
- Control over contaminants entering the system.

The use of microorganism-based cleaners offers a possible "in-process" microbial application of green chemistry and the process offers future potential in systems polluted with chlorinated and non-halogenated organic compounds.

156

Acknowledgements: WMN would like to express appreciation to IL Department of Military Affairs, ChemFree Corporation (Norcross, GA) and the Illinois Waste Management and Research Center for funding the work in this paper.

References

(1) Nelson, W. M. *Green Solvents for Chemistry: Perspectives and Practice*; Oxford University Press, Inc.: New York, 2003.

(2) Commerce, U. S. D. o. ""Cleaning up the nation's waste sites: markets and technology trends"," National Technical Information Service (NTIS), 1996.

(3) Anastas, P. T.; Warner, J. C. *Green Chemistry: Theory and Practice*; Oxford University Press: New York, 1998.

(4) Chapelle, F. H. *Groundwater microbiology and geochemistry*; John Wiley & Sons: New York, 1993.

(5) MacGillivray, A. R.; Shiaris, M. P. In *Biological Degradation and Bioremediation of Toxic Chemicals*; Chaudhry, G. R., Ed.; Dioscorides Press: Portland, OR, 1994, pp 125-147.

(6) Sutherland, S. S. *Remediation engineering design concepts*; Lewis Publishers: New York, 1997.

(7) Nelson, W. *Precision Cleaning* **1996**, *IV*, 30-34.

(8) Kanegsberg, B.; Kanegsberg, E., Eds. *Handbook for Critical Cleaning*; CRC Press LLC: New York, 2001.

(9) McNally, T. W. *Parts Cleaning* **1999**, 20-27.

(10) Ortiz, O.; Mc Nally, T. W. *CleanTech magazine* **2002**, *II*.

(11) Eskamani, G. "Evaluation of BioClean USA, LLC Biological Degreasing System for the Recycling of Alkaline Cleaners," CAMP, Inc., 2000.

(12) Osprey Biotechnics, 2003 <www.ospreybiotechnics.com/bioaugmentation.html>.

(13) Nelson, W. M.; Tkachenko, V. *Precision Cleaning* **2001**, 24-29.

(14) Weinstein, W. *Environmental Protection* **1996**, 28-29.

(15) Bonilla, E.; Bonilla, J.; Baker, M.; Smith, J. *CleanerTimes* **1995**, 21-22.

(16) Steffan, R. J.; McClay, K.; Vainberg, S.; Condee, C. W.; Zhang, D. *Appl Environ Microbiol* **1997**, *63*, 4216-4222.

(17) Harris, J. W. *Hydrocarbon Processes* **1991**, *64*, 154.

(18) Niemi, G. J. *Environ Toxicol Chem* **1987**, *6*, 156-171.

(19) Bustard, M. T.; McEvoy, E. M.; Goodwin, J. A. S.; Burgess, J. G.; Wright, P. C. *Appl Microbiol Biotechnol* **2000**, *54*, 424-431.

(20) Maymó-Gatell, X.; Chien, Y.-t.; Gossett, J. M.; Zinder, S. H. *Science* **1997**, *276*, 1568-1571.

(21) Cirpka, O. A.; Windfuhr, C.; Bisch, G.; Granzow, S.; Scholz-Muramatsu, H.; Kobus, H. *J.Env.Engineering* **1999**, *125*, 861-870.

(22) Aziz, C. E.; Georgiou, G.; Speitel Jr, G. E. *Biotechnol Bioeng* **1999**, *65*, 100-107.

(23) Lee, P.-K.; Ong, S. K.; Golchin, J.; Nelson, G. L. *Wat. Res.* **2001**, *35*, 3941-3949.

(24) Yeom, I. T.; Ghosh, M. M.; Cox, C. D. *Environ. Sci. Technol.* **1996**, *305*, 1589-1595.

(25) Kilbane, J. J. *Water Air Soil Pollut.* **1997**, *104*, 285-304.

Application of Green
Chemistry Principles
in Agriculture

Chapter 11

Management of the Soybean Cyst Nematode by Using a Biorational Strategy

George A. Kraus[1], Gregory L. Tylka[2], Steve Van der Louw[1], and Prabir K. Choudhury[1]

Departments of [1]Chemistry and [2]Plant Pathology, Iowa State University, Ames, IA 50011

Analogs of glycinoclepin A have been shown to affect hatching of soybean cyst nematode cysts in laboratory conditions. Most analogs are hatch inhibitors. More complex bicyclic compounds may be hatch accelerators.

Introduction

Soybean cyst nematode (SCN), also known as *Heterodera glycines*, is one of the most widely distributed and economically devastating soybean pests. SCN was first reported in Japan by Hori in 1915 and first appeared in the United States in North Carolina in 1954 (1). SCN has since been confirmed in 28 states. SCN was first detected in Iowa in 1978 in Winnebago County and currently the majority of Iowa counties are known to be infested with SCN. It can be assumed that undetected infestations are probably present in many other counties as well. Documenting the economic impact of SCN is difficult, because detection of SCN is difficult in the early stages and the producers may attribute the loss in yields to factors other than SCN. If nationwide loss is conservatively estimated at 1%, SCN costs soybean producers $121 million in 1992 alone. Typically, the estimated percentage losses are not 1%, but range from 1.1- 5.8%. An Iowa farm field ravaged by SCN is illustrated below in Figure 1. The light green, yellow, and brown areas are areas of significant SCN damage.

162

Figure 1. Soybean field infested with soybean cyst nematode.

SCN survives in the soil as eggs contained within protective cysts (2). Many of the eggs contain fully developed second-stage juveniles that will hatch under the proper conditions. The cyst typically contains 200 or more eggs. Because of the relatively short life cycle of the SCN (24 to 30 days), under the proper conditions, three to four generations of SCN can be produced in a single growing season. This ability of the SCN to rapidly proliferate, combined with its hardiness and longevity make the SCN difficult to control. The SCN is readily dispersed by the movement of infested soil through adherence to machinery, and may also be dispersed by wind, runoff water, livestock, wildlife, and migrating birds.

Currently, management of the SCN is achieved by incorporation of a crop rotation strategy, use of resistant soybean varieties, and the use of nematicides (3). Because of environmental and economic concerns, the use of nematicides has dwindled over the years. In place of nematicides, crop rotation strategies have been utilized where two years of non-host crops are used followed by one year of SCN resistant soybeans. This strategy has been effective in controlling SCN population densities, but market considerations make this strategy unattractive, if not unfeasible, to producers. Also, the overuse of SCN-resistant soybean varieties may lead to the development of SCN races which can readily reproduce on resistant soybeans.

In recent years there has been more interest in the possible development of herbicides that affect the hatch of the SCN (4,5). Herbicides which stimulate or inhibit the hatch could be used to manage the SCN populations if they could cause the SCN to hatch prematurely in the absence of a host plant or completely suppress the hatch when soybeans are being grown. In the absence of a host plant, the SCN would not be able to reproduce and would ultimately die from starvation, parasitism, or predation. If the SCN was found to be present during the growing season, a herbicide that suppressed hatch could be used to keep the population densities from proliferating.

Glycinoclepin A is a hatching stimulus capable of initiating hatch of SCN eggs at concentrations as low as 10^{-12} g/mL (6). It is a naturally occurring compound that should be readily biodegradable. Glycinoclepin A has been isolated by extraction from kidney bean roots, but only milligram quantities were obtained from thousands of kilograms of roots. The structure of glycinoeclepin A is shown in Figure 2.

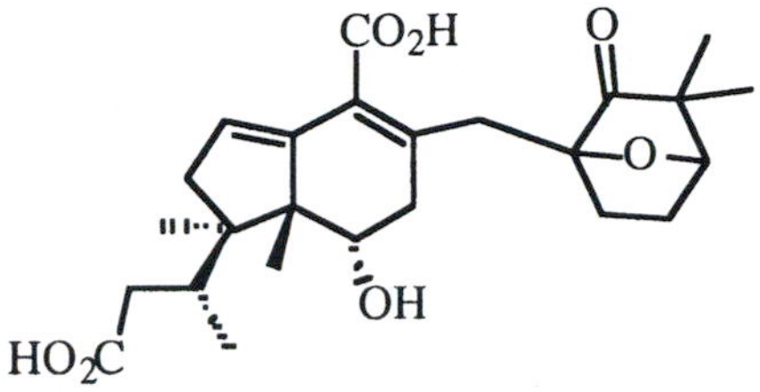

Figure 2. Structure of glycinoclepin A

Murai and coworkers set out to determine the minimum functionality needed in order to stimulate SCN hatch (7). They synthesized numerous compounds and measured the minimum concentration accelerating the hatching of the juveniles from half the number of eggs. It was determined from this study that the minimum functionality to induce hatch are the axial hydroxyl group and the two carboxylic acids. Of significant interest from a synthetic chemistry perspective, the side chain functionality in the oxabicyclic ring system and also the position of the cross-conjugated diene has little effect on the activity of the analogs. Kraus and Tylka also reported their studies on the functionality necessary for biological activity (8). A summary of the Mauai and Kraus findings is collated in Figure 3.

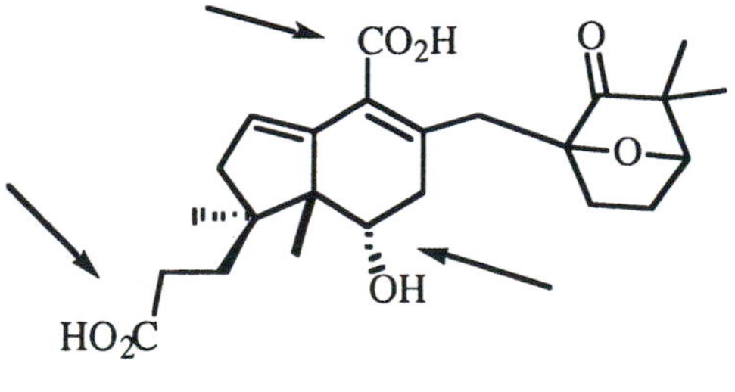

Figure 3. Key functional groups for bioactivity

The structure activity relationship information proved to be invaluable in determining the course of our research project. Corey synthesized glycinoeclepin A by a multistep route (9). Murai (10, 11) and Mori (12, 13) also synthesized glycinoeclepin A by multistep routes. Corey later published a more direct synthesis using a novel rearrangement reaction (14). Miwa and coworkers have reported a clever approach to a key fragment of glycinoeclepin A (15).

We wished to develop a series of small molecule activators that would allow us to imitate the activity of glycinoclepin A. Our overall goal was to find a route to a small molecule analog that would be practical and would be amenable to scale up for industrial production. These small-molecule activators should be synthetically accessible in the fewest number of steps possible. It was also crucial that the starting materials for the syntheses were inexpensive and readily available. With these goals in mind, we set out to discover new and potent compounds to control the proliferation of the soybean cyst nematode.

Results and Discussion

The synthesis of the analogs began with a plan to create a molecule that has as many of the necessary active groups as possible. Using literature methods (16, 17), we synthesized compounds **2-8**. Their structures are listed in Figure 4. These compounds were prepared from readily available starting materials such as ethyl acrylate, cyclohexanone, cyclopentanone and diethyl oxalate. Compound **2** was synthesized by the method of Stork (16) from the enamine of cyclohexanone and ethyl acrylate, followed by hydrolysis.

The compounds were tested using the chamber shown in Figure 5. He used distilled water as a control and zinc sulfate, a known hatch stimulator, as a standard for comparison. The results of the screening, shown in Figure 6, surprised us. Unlike glycinoclepin A and zinc sulfate, compounds **7** and **8** were not hatch stimulators but rather hatch suppressers. In addition, compounds **7** and **8** were also extremely potent, with **7** inhibiting the hatch at the 1 ppm level. It is also worth noting that if the compounds were washed away after 30 days and the eggs placed in water or zinc sulfate, the hatch profile resumed its normal curve. Therefore, clearly compounds **7** and **8** were not killing the SCN, but were repressing hatch by some unknown mechanism (19).

Realizing that our best hatch inhibitors contained an enolic diketone and wanting to test the effect of the six member ring upon activity, an acyclic and six

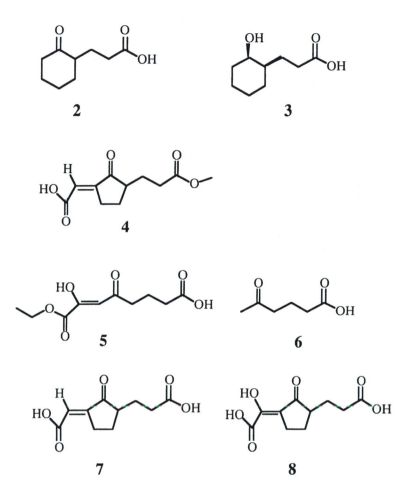

Figure 4. Initial array of compounds tested.

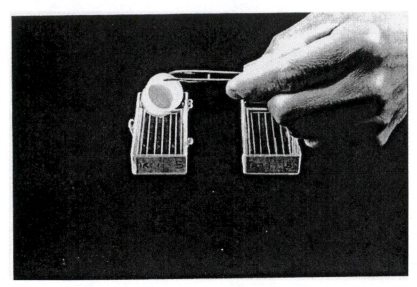

Figure 5. SCN hatch chambers

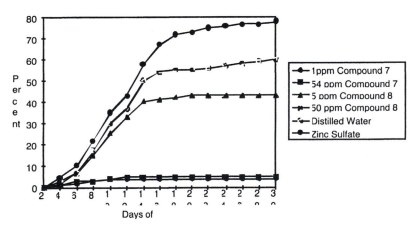

Figure 6. Results of testing of compounds 7 and 8.

membered ring analog were synthesized and tested. Compounds **9** and **10** (Figure 7) showed significant hatch suppression activity, but were not as effective as **7** and **8**, as a higher concentration of **9** and **10** were needed in order to achieve the same level of activity. Analyzing our results, we determined that of the compounds that are active in suppressing SCN hatch, they all had one of two common functional groups; the enolic dicarbonyl or the diacid.

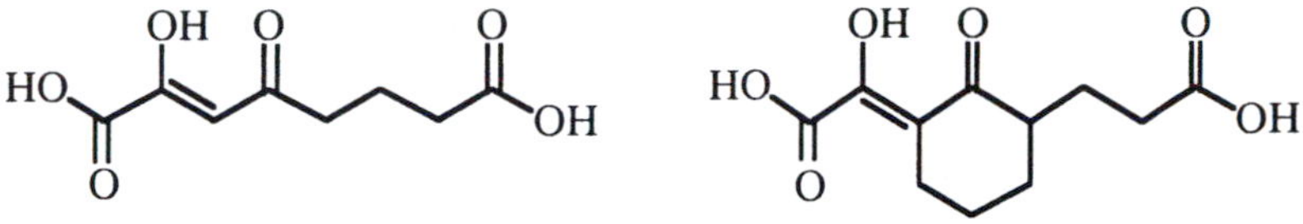

Figure 7. Compounds 9 and 10.

Compounds **11**, **12**, **13**, and **14** (Figure 8) were all submitted for testing and the results were encouraging. There was significant hatch suppression by **11** and **12**, but the difference between the two was minimal. This demonstrated to us that only the enolic dicarbonyl was necessary. Similar results were found with **13** and **14**. We can synthesize compounds in one step (from readily available and inexpensive starting materials) that effectively contained SCN hatch.

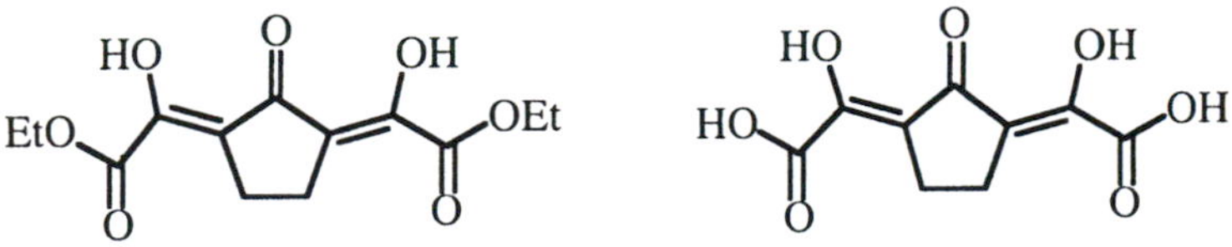

Figure 8. Compounds 11 to 14.

We felt it was worth our time to determine if ring size had anything to do with the activity. We synthesized the six and seven membered ring analogs shown in Figure 9 by literature methods and tested them for activity. The results were similar to the five member ring analogs, with the ester being just as active as the acid, but overall the larger ring size seems to diminish activity slightly.

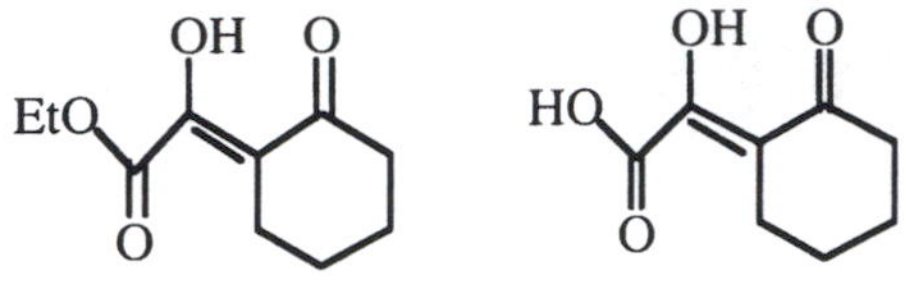

15   **16**

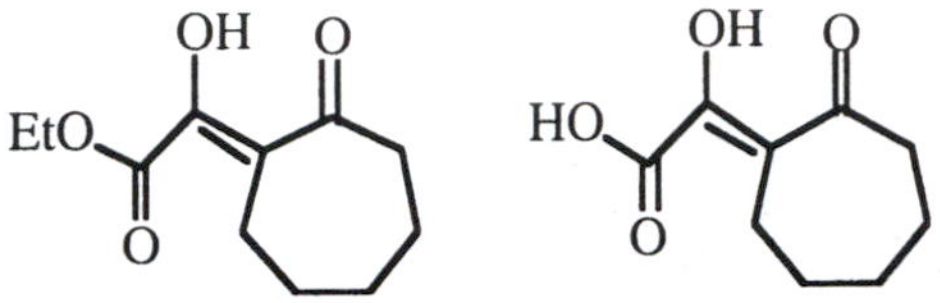

17   **18**

Figure 9. Six- and seven-membered ring analogs

The final set of experiments to determine the minimum functionality required for hatch inhibition were centered around determining if the ester group of the enolic dicarbonyl was necessary for activity or if a simple enolic carbonyl was just as effective. To accomplish this, we synthesized the enolic derivative **19** shown in Figure 10 (20). This compound was as active as any other analog in hatch inhibition. This implies that the active portion of the molecule is centered around the interaction of the enol with the SCN (cyclopentanone had no activity). The addition of the side chain ester or acid offered no enhancement of activity.

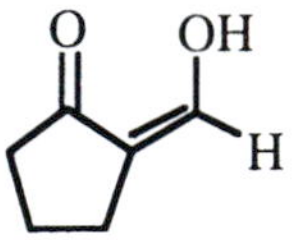

19

Figure 10. Hydroxymethylenecyclopentanone

As the synthetic effort to make efficient inhibitors continued, we also studied new synthetic transformations to afford more complex molecules that would act as hatch initiators. As part of this research plan, we developed two new synthetic methods that would accelerate our ability to make new bicyclic compounds (21, 22). These are shown in Figure 11.

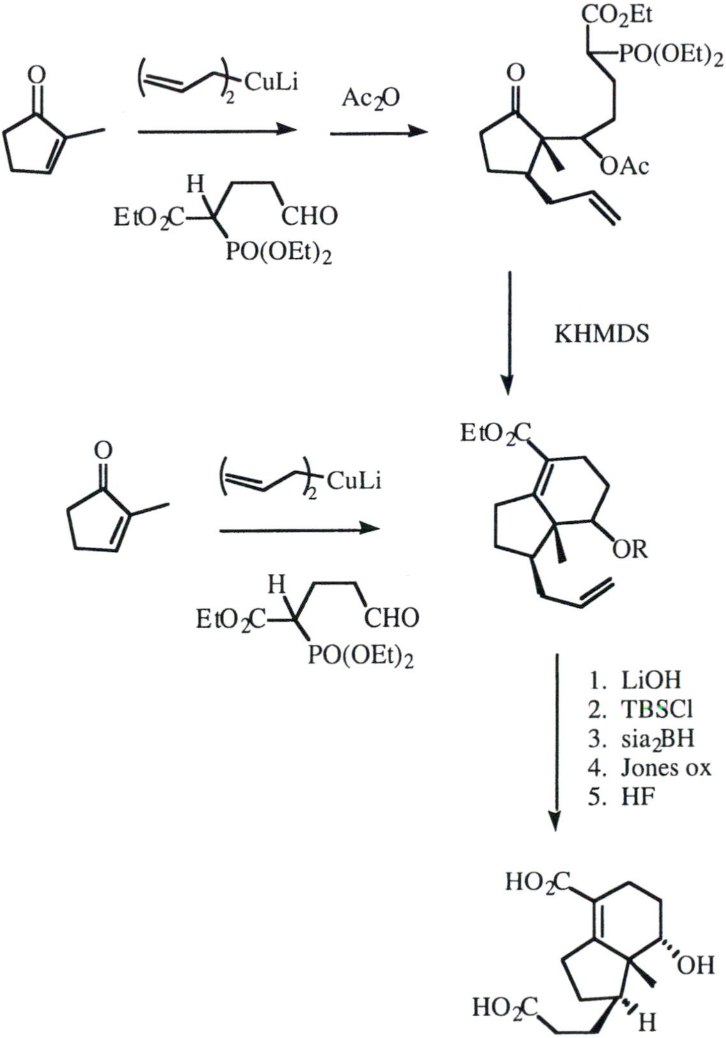

Figure 11. New annulation method for synthesis of glycinoeclepin A analogs

The first version of our annulation involved an aldol reaction with a phosphonate aldehyde followed by acetylation and an intramolecular Emmons cyclization. This three-step procedure afforded the bicyclic ring system shown above in approximately 20% overall yield from 2-methylcyclohexenone. We later identified reaction conditions that enabled us to achieve the same transformation in *one pot* in over 30% yield. The resulting bicyclic ester was converted into the acid by ester hydrolysis with LiOH. Protection of the alcohol as the tert-butyldimethylsilyl ether, elaboration of the allyl group into the propionic acid side chain and removal of the protecting group provided the desired diacid. This compound will be tested for inhibitory activity against SCN. The new annulation will permit us to easily synthesize a variety of more complex analogs.

Conclusions

Although our idea when we began this study was to develop small molecules that cause premature hatch of the SCN, we have synthesized nine compounds that effectively inhibit the hatch of the soybean cyst nematode. During the course of the study we have determined that the minimum functionality for activity is the enolic dicarbonyl. The five membered ring analogs are more effective than the six and seven membered ring analogs at inhibiting hatch. We have also developed an annulation reaction which will allow us to easily prepare more complex bicyclic analogs.

References

1. Winstead, N.N., C.B. Skotland, and J.N. Sasser "Soybean-cyst nematode in North Carolina" *Plant Disease Reporter* 1955, 39, 9-11.
2. Niblack, T. L., N. K. Baker, D. C. Norton "Soybean yield losses due to *Heterodera glycines* in Iowa" *Plant Disease* 1992, 76, 943-948.
3. Tylka, G.L. "Soybean Cyst Nematode" *Iowa State University Extension publication Pm-879* 1995, 6pp
4. Kraus, G. A.; Johnston, B; Kongsjahju, A.; Tylka, G. L. "The Synthesis and Evaluation of Compounds that Affect Soybean Cyst Nematode Egg Hatch" *J. Ag. Food Chem.*, **1994**, *42*, 1839.
5. Vander Louw, Steven John. Synthesis of biomimetic compounds (natural antenna, soybean cyst nematode, hatch inhibitors, glycinoeclepin A). (1997), 102 pp. CAN 127:346539

6. Masamune, T., M. Anetai, M. Takasugi, N. Katsui "Isolation of a natural hatching stimulus glycinoeclepin A for the soybean cyst nematode" *Nature* **1982**, 297, 495-7.

7. Murai, A., M. Ohkita, T. Honma, K. Hoshi, N. Tanimoto, S. Araki, A. Fukuzawa "Structure-activity relationship of glycinoeclepin A" *Chem Letters*, **1992**, 2103-4

8. Kraus, G. A.; Vander Louw, S.; Tylka, G. L.; Soh, D. H. "The Synthesis and Testing of Compounds that Inhibit Soybean Cyst Nematode Egg Hatch" *J. Agric. and Food Chem.*, **1996**, *44*, 1548.

9. Corey, E. J., I. N. Houpis "Total synthesis of glycinoeclepin A" *J. Am. Chem. Soc.* **1990**, 112, 8997-9 and references therein.

10. Murai, Akio; Tanimoto, Norihiko; Sakamoto, Noriyasu; Masamune, Tadashi. "Total synthesis of glycinoeclepin A". *J. Am. Chem. Soc.* **1988**, 110, 1985-6

11. Murai, Akio. "Total synthesis of glycinoeclepin A". *Pure Appl. Chem.* **1989**, 61, 393-6

12. Mori, Kenji; Watanabe, Hidenori. "Recent results in the synthesis of semiochemicals: synthesis of glycinoeclepin A". *Pure Appl. Chem.* **1989**, 61, 543-6.

13. Watanabe, Hidenori; Mori, Kenji. "Triterpenoid total synthesis. Part 2. Synthesis of glycinoeclepin A, a potent hatching stimulus for the soybean cyst nematode" *J. Chem. Soc., Perkin Trans. 1* **1991**, 2919-34.

14. Corey, E. J.; Hong, B. "Chemical Emulation of the Biosynthetic Route to Glycinoeclepin from a Cycloartenol Derivative" *J. Am. Chem. Soc.* **1994**, 116, 3149-50.

15. Miwa, Atsushi; Nii, Yasushi; Okawara, Hideki; Sakakibara, Masayuki. "Synthetic study on hatching stimuli for the soybean cyst nematode" *Agric. Biol. Chem.* **1987**, 51, 3459-61.

16. Stork, G., A. Brizzolara, H. Landesman, J. Szmuszkovicz, R. Terrell "The enamine alkylation and acylation of carbonyl compounds" *J. Am. Chem. Soc.*, **1963**, *85*, 207.

17. Bonadies, F., Scarpati, M. L. "Condensation of α, γ diketo (or keto succinic) esters with gyloxalic acid" *Gazzetta Chimica Italiana*, **1983**, *113*, 421-425.

18. Wong, A. T. S., G. L. Tylka, R. G. Hartzler "Effects of eight herbicides on in vitro hatching of *Heterodera glycines*" *J. Nematology*, 1993, 25, 578-584.

19. S. M. Pike, R. Heinz, T. Walk, C. Jones, G. A. Kraus, A. J. Novacky, T. L. Niblack "Is Change in Electrical Potential or pH a Hatching Signal for *Heterodera glycines*?" *Phytopathology*, **2002**, *92*, 451.

20. Eaton, P. E.; Jobe, P. G. "Improved syntheses of hydroxymethylene cyclopentanone and spiro{4.5}dec-6-ene-1,8-dione" *Synthesis*, **1983**, 796-797.

21. G. A. Kraus, C. Jones "The Reaction of Ketone Enolates with a δ-Oxo Phosphonate: A Carbanion-Based 4+2 Annulation" *Synlett*, **2001**, 793.
22. G. A. Kraus, P. K. Choudhury "Phosphonate Aldehyde Annulation. A One-Pot Synthesis of δ-Hydroxy Cyclopentenoic Esters", *Organic Letters*, **2002**, *4*, 2033.

Chapter 12

Potential of Entomopathogenic Fungi as Biological Control Agents against the Formosan Subterranean Termite

M. S. Wright, W. L. A. Osbrink, and A. R. Lax

Formosan Subterranean Termite Research Unit, SRRC, Agricultural Research Service, U.S. Department of Agriculture, 1100 Robert E. Lee Boulevard, New Orleans, LA 70124

Tolerance, pathogenicity and transmission studies of the fungi *Metarhizium* and *Beauveria*, show that biological control agents can enhance termite treatment flexibility. Subterranean termites cause significant damage to wood structures and trees, especially along the Gulf of Mexico coastal region of the United States. A predominant pest species is the Formosan subterranean termite, *Coptotermes formosanus* (Shiraki), which differs from native termite species in increased colony density and a propensity to destroy living wood. However, in order for termite control approaches to work they must be non-repellant, transferrable, and have delayed toxicity to allow transfer from foraging workers to their nestmates. An Integrated Pest Management (IPM) approach will be necessary to reduce the impact of these pests. One component of IPM, and the focus of this work, is the development of biological control agents. Environmental conditions in FST nests and sites of infestation, such as living trees, can vary greatly. Some treatment sites require novel treatment methods which fungi may be uniquely suited to provide.

Introduction

Subterranean termites are wood-destroying pest insects that cause structural instability in cellulose-based structures such as wood buildings and trees. They result in approximately $1 billion in damage and repair costs in the United States annually (*1*). One species, *Coptotermes formosanus* Shiraki, the Formosan subterranean termite (FST), is responsible for a significant amount of the overall impact. FST differ from other subterranean termites in their ability to survive in secondary nests with no contact to the original underground nest, propensity to attack living trees, and ability to maintain larger colonies than those of native subterranean termites. FST are believed to have entered the US mainland post-World War II (*2*). They have since become well established in coastal regions such as Galveston, TX, Lake Charles, LA and Charleston, SC. One location that has experienced the establishment of large, extensive FST colonies is New Orleans, LA where FST have caused damage in the historic French Quarter as well as in other neighborhoods and in the city's numerous trees (*2*). In 1998, Operation Full Stop, a comprehensive program to develop techniques to control the impact of FST in the US was established at the Southern Regional Research Center in New Orleans, LA. Operation Full Stop is charged with demonstrating the most effective means of use for currently available termite management tools, such as baits and non-repellant termiticides, to develop novel methods for termite control, and to communicate the information to the pest control industry and property owners. Biological control is a valuable component of an Integrated Pest Management scheme that also includes chemical and physical control. Biological control agents can offer a treatment option if a property owner prefers not to use chemicals in their home, when living trees or plants are infested by termites, or in combination with chemicals to effect synergism (*3, 4*). Some characteristics of termite nests and termite behavior, such as the presence of antifungal chemicals and the avoidance by FST of some pathogens and ailing nestmates, have the potential to limit the effectiveness of biological control agents (*5, 6, 7*). It is therefore important to select microorganisms that exhibit specific traits such as non-repellency, tolerance of termite nest conditions and delayed action to optimize the potential effectiveness. In this work fungi have been investigated as control options against FST (*8*). Microorganisms were observed for effectiveness against termites based on two criteria. Those that were known to infect other insects were screened for their pathogenicity against FST and tolerance of termite nest conditions. Those microorganisms that were known to co-exist with termites in their natural habitat, and presumably tolerated nest conditions, were screened for pathogenicity against FST. Microorganisms were isolated from the soil, ailing and dead termites, carton nest material and culture collections.

Materials and Methods

Isolation and Propagation of Fungi

Curvularia lunata was provided by Dr. Alesia N. Parker of the USDA/ARS/SRRC (New Orleans, LA) and was plated onto potato dextrose agar (PDA) and incubated at 25 C for 14 days. *Metarhizium anisopliae* strain ESC-I was provided in a commercial powder formulation (BioBlast, EcoScience, Worcester, Massachusetts) by Dr. Gregg Henderson of the Louisiana State University Agricultural Center (Baton Rouge, LA). *Beauveria bassiana* strains were obtained from the American Type Culture Collection (ATCC). Because none of the strains were previously known to pathogenize termites they were chosen based on variability of their isolation sites. Strains 26037 and 90519 were isolated from beetles in Colorado and Japan, respectively. Strain 90518 was isolated from Oregon soil. Pure cultures of the fungi were grown on PDA plates that were incubated at 25 C for 7-14 days, then stored at 4 C prior to use.

Exposure of Fungi to Volatile Chemicals

Spores were harvested by adding 10 mL of 0.01% Triton X-100 (Amresco, Solon, Ohio) to *B. bassiana* and *M. anisopliae* culture plates, scraping the surface of the plates with inoculating loops, and transferring each suspension to a sterile centrifuge tube. Spore suspensions, as determined using a Levy hemacytometer, were diluted to a concentration of 1.9×10^7 with 0.01% Triton X-100. PDA was poured into Petri dishes and allowed to harden. A sterile 10 mm cork borer was then used to core one hole in the center of the agar and another at the edge, and a sterile inverted septum cap was placed in the hole at the edge of the plate (9). A 500 µL aliquot of a fungal suspension of either *B. bassiana* or *M. anisopliae* was pipetted into the center well of a cored plate. Volatile chemicals that have been identified in association with termite nests were examined for their ability to inhibit fungal growth. Naphthalene (Fisher, Fair Lawn, New Jersey), (+) fenchone (98+%, Lancaster, Morecambe, England), and/or (-) fenchone (98+%, Lancaster, Morecambe, England) was pipetted into the inverted septum cap at the edge of an inoculated plate. Each plate was sealed with parafilm and plates having the same concentration of each volatile compound were incubated in a 4.5 L capacity modular chamber at 26 C to minimize reduction in concentration due to diffusion. Control plates were inoculated but did not contain a volatile chemical. Treatments and the control were replicated three times. At each evaluation the radial growth of the culture, from the edge of the center well to the edge of the hyphal growth, was measured.

Collection of Termites

Formosan subterranean termites were collected in bucket traps on the campuses of the University of New Orleans and the Southern Regional Research Center in New Orleans, Louisiana following the method of Su and Scheffrahn (10). Four colonies were selected on each campus with one colony representing one replicate within an experiment. Each replicate contained 10 to 20 workers of at least 3rd instar, as determined by size. Termites were used within one week of field collection.

Determination of Fungal Pathogenicity

Cultured agar plates of each fungus stored at 4 C were allowed to reach room temperature prior to termite exposure. For each of the four replicates per experiment 10 or 20 workers were placed on the fungal culture plate for 10 minutes to allow attachment of the fungal spores to the termite cuticle. The workers were then transferred to sterile moistened #4 filter paper (Whatman, Maidstone, England) in a 100 X 15 mm Petri dish (Falcon, Franklin Lakes, New Jersey). Control plates contained sterile moistened filter paper and termites that were not exposed to fungi. All plates were incubated in the dark at 26 C and 99% humidity for the duration of the experiment. The number of dead termites was recorded at each time point.

Determination of Fungal Disease Transmission

Cultured agar plates stored at 4 C were allowed to reach room temperature prior to termite exposure. For each of the four replicates per experiment 10 workers were placed on the fungal culture plate for 10 minutes to allow attachment of the fungal spores to the termite cuticle. The workers were then transferred to sterile moistened #4 filter paper (Whatman, Maidstone, England) in a 100 X 15 mm Petri dish (Falcon, Franklin Lakes, New Jersey) which also contained 10 of their nestmates that had not been exposed to the fungus. Control plates contained sterile moistened filter paper and termites that were not exposed to a fungal culture. All plates were incubated in the dark at 26 C and 99% humidity for the duration of the experiment. The number of dead termites was recorded at each time point.

Recovery of Microbes from Termite Cadavers

At each observation dead termites were removed from the experimental plate and were placed onto a PDA plate to allow growth of associated fungi. The cadaver plates were incubated at 25 C until fungal sporulation was visible, generally at 7 days. The presence or absence of the applied organism, *B. bassiana* or *M. anisopliae*, was recorded for each cadaver. In many cases an abundance of bacterial cultures was also observed and recorded. One bacterial species commonly found in association with termites, *Serratia marcesans*, was apparent by its red pigmentation and was noted when present.

Modification of Fungal Repellency

Spores of *M. anisopliae* strains 1186, and 1218 were provided by Dr. Richard Milner (CSIRO, Canberra, Australia). Twenty FST workers were exposed to the fungal spores individually or in combination by placing the termites in a Petri dish that contained a total of 0.2 g of spores on filter paper moistened with 1.5 mL of sterile water. To measure the potential of one strain to modify the repellency of the other, FST workers were exposed to the following spore mixtures on moistened filter paper: 100% 1186; 70% 1186 + 30% 1218; 50% 1186 + 50% 1218; 30% 1186 + 70% 1218; 100% 1218. The experiment was replicated 4 times for each combination of fungal spores. Petri dishes were incubated at 26 C for the duration of the experiment. The number of dead termites was counted at each time point.

Results and Discussion

Exposure of Fungi to Volatile Chemicals

Fungi such as *B. bassiana* and *M. anisopliae* display many of the beneficial characteristics of successful termiticidal chemicals, such as delayed action and the ability to be spread by termite social behavior including trophallaxis and allogrooming (*11*). In addition, fungi are self-replicating and the environmental conditions within a termite nest can be conducive to fungal growth. Fungi can potentially infect colony members after growing on ailing or dead nestmates, and/or on carton nest material (*12, 13*). The pathogenic potential of fungi can be aided by their production of destruxins which are insecticidal chemicals (*14*). Subterranean termites survive in an environment that is rich with microbial growth and without control measures termite colonies would be overwhelmed. One of the factors that can contribute to the insect/microbe balance is the presence in the nest of antifungal volatile chemicals. Two chemicals in particular, naphthalene and fenchone, have been identified in termite nests (*15, 16*) One objective of this study was to determine the potential for synergistic inhibition by these chemicals on growth of the entomopathogenic fungi *M. anisopliae* and *B. bassiana*. Due to the difficulty of measuring the concentration of volatile chemicals in a closed termite nest, a range of inhibition for each chemical and each fungus was determined in the laboratory. Chemical concentrations that resulted in intermediate levels of inhibition, 20-60%, were selected for the synergism studies (*17*). Combined exposure of *M. anisopliae* strain ESC-1 to 1 mg naphthalene plus 10 μL (+) fenchone (62.3% of control growth), or to 1 mg naphthalene plus 10 μL (-) fenchone (73.6%) did not significantly increase inhibition over the level caused by exposure to either 1 mg naphthalene (68.7%), 10 μL (+) fenchone (77.4%), or 10 μL (-) fenchone (71.7%) alone (Figure 1). These data are from Wright et al. 2000 (*17*). A synergistic effect was also not observed when *B. bassiana* strain 26037 was exposed to combinations of the volatile chemicals. With *B. bassiana*, intermediate levels of inhibition were observed with 0.1 mg naphthalene and 1 μL fenchone. Exposure of *B. bassiana* to naphthalene plus (-) fenchone (53.8%), to naphthalene plus (+) fenchone (49.1%), or to naphthalene plus both (-) fenchone and (+) fenchone (53.8%) did not significantly increase inhibition over that of each volatile chemical alone (Figure 2). While all of the volatile chemicals did inhibit fungal radial growth, a primary concern was the potential for a combination of volatile chemicals to completely inhibit the fungi, especially considering the closed microclimate of a termite nest (*18*). These results indicated that *M. anisopliae* strain ESC-1 and *B. bassiana* strain 26037 were not completely inhibited by exposure to combinations of these compounds in a simulated closed environment, and further evaluation of their pathogenic potential was undertaken.

Determination of Fungal Pathogenicity and Transmissibility

FST workers from four colonies were exposed to spores of *C. lunata* and allowed to feed on moistened filter paper with an equal number of their nestmates. The mortality rate of exposed termites (53.8%) did not differ from that of unexposed control termites (53.8%) over the experimental period of 11 days (Figure 3). Furthermore, because the mortality rate only slightly exceeded 50% this may indicate that only termites that were directly exposed to the fungus were

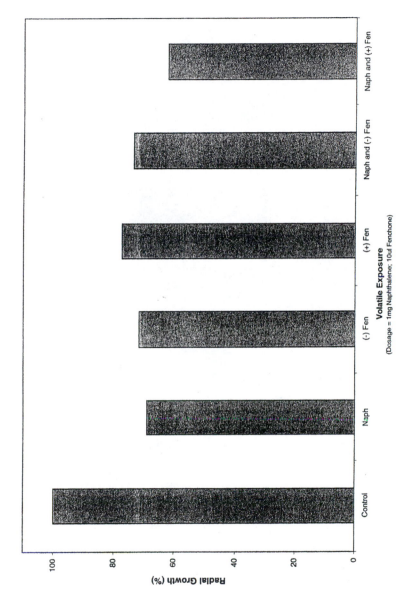

Figure 1: Radial Growth of *M. anisopliae* in the presence of termite nest volatile chemicals.

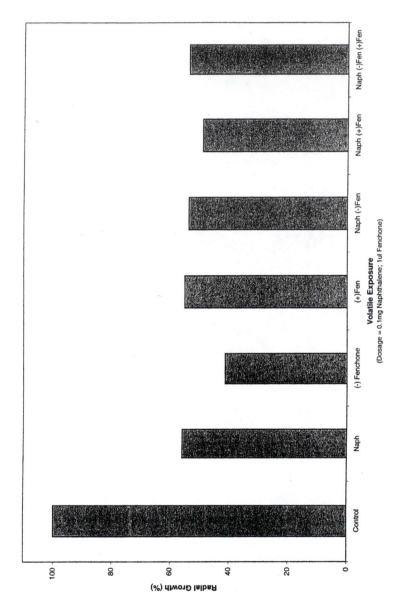

Figure 2: Radial Growth of *B. bassiana* in the presence of termite nest volatile chemicals.

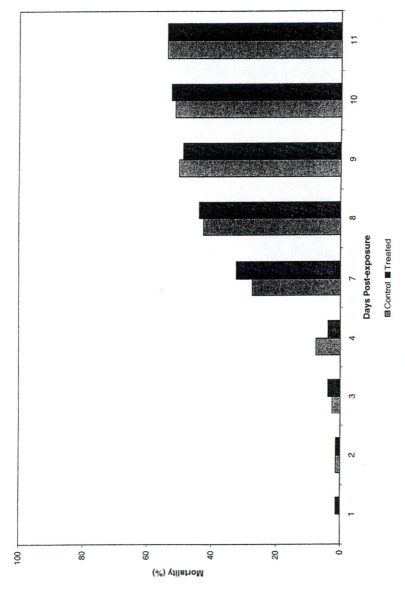

Figure 3: Pathogenicity of FST workers exposed to *C. lunata* either directly or indirectly.

pathogenized by it. Because of its low rate of mortality this strain has not been pursued further. An alternate approach for the isolation of fungal biological control agents involved the screening of microbes that are known to infect other insects. Fungi have unique potential to serve in this capacity because they commonly inhabit soil, and often have a preference for environmental conditions such as those found in termite nests. Their ability to infect termites using cuticle degrading enzymes may also afford them an advantage over bacterial biological control agents which must be ingested in order to kill termites (19). B. bassiana and M. anisopliae have previously been shown to infect subterranean termite species in the genera Coptotermes and Reticulitermes (20, 21, 22). FST workers from four colonies were exposed to spores of either B. bassiana or M. anisopliae and were then transferred to moistened filter paper with an equal number of their nestmates. Mortality of the group was monitored for 21 days. At day 7, two of the B. bassiana strains, 26037 and 90519, had caused 100% mortality among all 4 colonies (Figure 4). M. anisopliae strain Ma ESC-1 killed 68.8% of the termites by day 7 and 100% by day 15. These data indicate that the fungi killed not only all of the directly exposed termites, but also all of those that were subsequently infected through contact with their nestmates. Because only certain colony members forage outside of the nest area, transmission between nestmates will be critical to the success of any biological control agent.

Recovery of Microbes from Termite Cadavers

Termite mortality in excess of 50% presumably indicated that the experimental fungus was transferred from the workers that were allowed to walk in the fungal culture to their nestmates. To further confirm transmission, all dead termites were removed to PDA plates and observed for the presence of either M. anisopliae or B. bassiana. The percentage of cadavers on which each fungus grew was averaged for the four experimental colonies. B. bassiana strains 26037 and 90519 yielded fungal recovery of 85% and 74%, respectively (Figure 5). Among termites exposed to B. bassiana strain 90518 the fungus was recovered from only 32% of the cadavers. It is interesting to note that of the B. bassiana strains 90518 also caused the lowest rate of mortality of 42.5% at day 7 (Figure 4). M. anisopliae strain ESC 1, while causing 100% mortality at day 15 (Figure 4), was retrieved from an average of only 49% of termite cadavers (Figure 5). These results may reflect overgrowth of some M. anisopliae colonies by other microbes that were already on the termites or a variation in termite colony sensitivity to the pathogen. As shown previously by Wright, et al., among the colonies exposed to M. anisopliae ESC 1 recovery ranged from 35% to 75% (8). Neither B. bassiana nor M. anisopliae were recovered from the total of 11 control cadavers that were not exposed to the fungi.

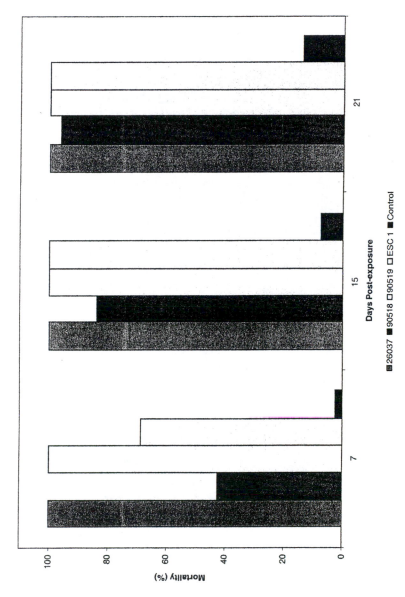

Figure 4: Pathogenicity of FST workers exposed to three strains of *B. bassiana* and one strain of *M. anisopliae.*

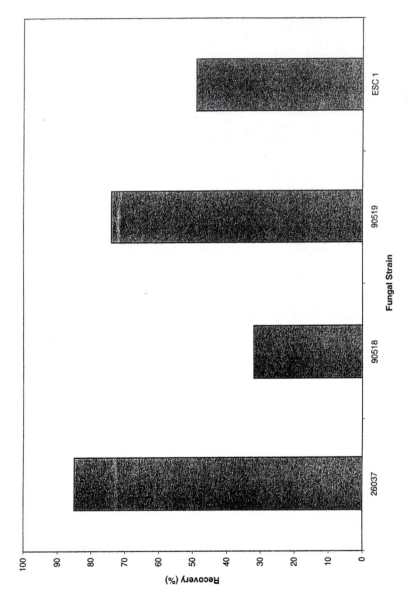

Figure 5: Recovery of *B. bassiana* or *M. anisopliae* from the cadavers of FST workers exposed either directly or indirectly.

Modification of Fungal Repellency

Previous studies have demonstrated that different strains within the same species can vary widely in their virulence potential (*23*). This phenomenon is also demonstrated in this study when comparing *B. bassiana* strains 26037 and 90519 to 90518. This variability may reflect chemical changes from strain to strain that result in termite repellency by some strains but not others. If termites are repelled, the fungus will not have the opportunity to attach to the cuticle of the exposed termite or will not be further transferred throughout the colony. To evaluate the ability of one strain to mask repellency of another, spores of two *M. anisopliae* strains were mixed and variations in mortality rates were determined. Previous work by Dr. Richard Milner showed that when FST workers were placed in a Petri dish with *M. anisopliae* strain 1186 they would walk through the spores, and spread the spores over the surface of the plate (*24*). When FST workers from the same colony were placed in a Petri dish with *M. anisopliae* strain 1218 they would avoid the area of the plate on which the spores were placed. The spores remained undisturbed and did not attach to the cuticle of the insects (*24*). In the present study FST workers were placed in Petri dishes with spores of either strain 1186 or 1218, or with one of the spore mixtures. All termites exposed to *M. anisopliae* exhibited higher mortality than the controls which were exposed to filter paper only. Strain 1186 had a rate of 67.5% at day 3 and reached 100% mortality by day 7. Conversely, strain 1218 had a rate of only 10.3% at day three and reached a maximal rate of 79.5% by day 10 (Figure 6). The mixture containing 70% 1186 spores paralleled the mortality rate of the 100% 1186 spores throughout the study. This result shows a degree of masking of the repellency caused by the 1218 spores. The mixtures containing either 50% or 30% 1186 also showed an improvement over the spores of strain 1218 alone, with maximal mortality rates of 94% and 83% at day 10, respectively (Figure 6). These data confirm that while a single fungal strain may not provide all of the ideal characteristics such as attachment, invasiveness, cuticle-degrading enzyme production, or lack of repellency, a mixture of spores from strains with unique advantages may provide enhanced effectiveness. The studies presented here demonstrate the potential of the fungi *B. bassiana* and *M. anisopliae* to effect control of FST populations due to their ability tolerate termite nest conditions, pathogenize termites and, when spores are combined, to mask repellency by a strain that has other advantageous features.

References

1. Su, N.-Y.; Scheffrahn, R.H. Economically important termites in the United States and their control. Sociobiology 1990, 17, 77-94.

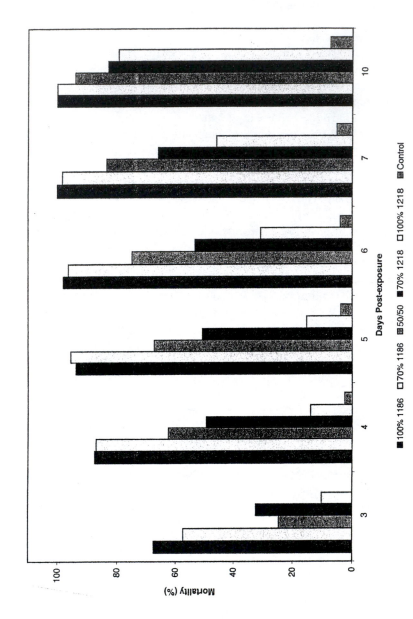

Figure 6: Mortality of FST workers exposed to mixtures of repellent and non-repellent strains of *M. anisopliae*.

2. LaFage, J. Practical considerations of the Formosan subterranean termites in Louisiana: a 30 year old problem. In Tamashiro, M.; Su, N.-Y. Biology and control of the Formosan subterranean termite, University of Hawaii, Honolulu, HI, 1987, 37-47.

3. Grace, J.K. Biological control strategies for suppression of termites. J. Agric. Entomol. 1997, 14, 281-289.

4. Culliney, T.W.; Grace, J.K. Prospects for the biological control of subterranean termites (Isoptera: Rhinotermitidae), with special reference to *Coptotermes formosanus*. Bull. Entomol. Res. 2000, 90, 9-21.

5. Wood, T.G.; Sands, W.A. The role of termites in ecosystems. In Brian, M.V. (ed.) Production ecology of ants and termites. Cambridge University Press, Cambridge, MA, 1978, 245-292.

6. Su, N.-Y.; Tamashiro, M.; Yates, J.R.; Haverty, M.I. Effect of behaviour on the evaluation of insecticides for prevention of or remedial control of the Formosan subterranean termite. J. Econ. Entomol. 1982, 75, 188-193.

7. Grace, J.K. Biological control strategies for suppression of termites. J. Agric. Entomol. 1997, 14, 281-289.

8. Wright, M.S.; Osbrink, W.L.A., Lax, A.R. Transfer of entomopathogenic fungi among Formosan subterranean termites and subsequent mortality. J. Appl. Ent. 2002, 126, 20-23.

9. Zeringue, H.J., Jr.; McCormick, S.P. Aflatoxin production in cultures of *Aspergillus flavus* incubated in atmospheres containing selected cotton leaf-derived volatiles. Toxicon 1990, 28, 445-448.

10. Su, N.-Y.; Scheffrahn, R.H. A method to access, trap and monitor field populations of the Formosan subterranean termite (Isoptera: Rhinotermitidae) in the urban environment. Sociobiology 1986, 12, 299-304.

11. Grace, J.K.; Zoberi, M.H. Experimental evidence for transmission of *Beauveria bassiana* by *Reticulitermes flavipes* workers (Isoptera: Rhinotermitidae). Sociobiology 1992, 20, 23-28.

12. Kramm, K.R.; West, D.F.; Rockenbaugh, P.G. Termite pathogens: transfer of the entomopathogen *Metarhizium anisopliae* between *Reticulitermes* sp. termites. J. Invertebr. Pathol. 1982, 39, 1-5.

13. Ignoffo, C.M. Environmental factors affecting persistence of entomopathogens. Florida Entomol. 1992, 75, 516-525.

14. Wahlman, M.; Davidson, B.S. New destruxins from the entomopathogenic fungus *Metarhizium anisopliae*. J. Nat. Prod. 1993, 56, 643-647.

15. Chen, J; Henderson, G.; Grimm, C.C.; Lloyd, S.W.; Laine, R.A. Naphthalene in Formosan subterranean termite carton nests. J. Agric. Food Chem. 1998, 46, 2337-2339.

16. Henderson, G. Personal communication.
17. Wright, M.S.; Lax, A.R.; Henderson, G.; Chen J. Growth response of *Metarhizium anisopliae* to two Formosan subterranean termite nest volatiles, naphthalene and fenchone. Mycologia 2000, 92(1), 42-45.
18. Wiltz, B.A.; Henderson, G.; Chen, J. Effect of naphthalene, butylates hydroxytoluene, dioctyl phthalate and adipic dioctyl ester, chemicals found in the nests of the Formosan subterranean termite (Isoptera:Rhinotermitidae) on a saprophytic *Mucor* sp. (Zygomycetes: Mucorales). Environ. Entomol. 1998, 27, 936-940.
19. Gupta, S.C.; Leathers, T.D.; El-Sayed, N.; Ignoffo, C.M. Insect cuticle-degrading enzymes from the entomogenous fungus *Beauveria bassiana*. Exp. Mycol. 1992, 16, 132-137.
20. Bao, L.L.; Yendol, W.G. Infection of the Eastern subterranean termite, *Reticulitermes flavipes* (Kollar) with the fungus *Beauveria bassiana* (Balsamo) Vuill. Entomophaga 1971, 16, 343-352.
21. Suzuki, K. Biological control of termites by pathogenic fungi. Conference Forestry and Forest Prod. Res. 1995, 146-156.
22. Delate, K.M.; Grace, J.K.; Tome, C.H.M. Potential use of pathogenic fungi in baits to control the Formosan subterranean termite. J. Appl. Entomol. 1995, 119, 429-433.
23. Wells, J.D.; Fuxa, J.R.; Henderson, G. Virulence of four pathogens to *Coptotermes formosanus* (Isoptera: Rhinotermitidae). J. Entomol. Sci.1995, 30, 208-215.
24. Milner, R.J. Personal communication.

Acknowledgments

The authors wish to thank Bridgette H. Duplantis, Erin E. Courtney and Amelia B. Mims for excellent technical assistance; and Anthony J. DeLucca II, Dr. Mary L. Cornelius, Dr. Alesia N. Parker and Dr. Hurley S. Shepherd for critical review of the manuscript.

Indexes

Author Index

Subject Index

198

essential oil-based pesticides, 46
pest management, 42
Environmental Protection Agency
(EPA)
commercialization of botanical
insecticides, 47
pesticide management, 12
reduced-risk pesticides, 42
Enzymatic activity,
protochlorophyllide oxidoreductase
C (POR C), 113
Enzymatic microbial degradation
aerobic metabolism, 155
anaerobic metabolism, 154
beyond industrial cleaning, 153–
154
BioClean Separator Module I,
146*f*
biodegradation of hydrocarbons,
142–145
ChemFree SmartWasher, 145*f*, 147,
150–151
chlorinated solvent remediation,
154–155
cleaning technology, 143–145
comfort level of workers, 150–
151
correlation between the colony
forming units (cfu) levels and
levels of fats, oils and grease
(FOG), 147, 150
fats, oils, and grease measurements,
152*f*
good cleaning performance, 150
limitations of aqueous cleaning,
146
microbiology, 144–145
microorganisms, 146–147
Mineral Masters bioremediating
technology, 151, 153
number of colony forming units in
Ozziejuice® samples, 148*t*
oil and grease levels in
Ozziejuice®, 149*t*
PAH biodegradation, 155

Enzymes
checking purity with marker, 111
See also Enzymatic microbial
degradation
Enzyme technology
economics, 12
green chemistry, 11–12
Erosion, soil degradation, 16–17
Escherichia coli
ethanol from distillers dried grains
with solubles (DDGS), 65–66
expression of *Arabidopsis*
chlorophyll *a* oxygenase, 32–
34
fiber fermentations with strain
FBR5, 75*t*, 76
pretreatment and fermentation of
distillers wet grains (DWG), 71,
73
See also Dry grind process
Essential oils
commercialization, 47–49
constituents with insecticidal and
fungicidal activities, 44*f*
neurotoxic mode-of-action, 46
persistence, 47
pesticides, 43–47
plant, with pesticidal actions in
insects/fungi, 43*t*
toxicity, 45*f*
See also Green pesticides
Ethanol
dry grind process, 64–65
production, 64
See also Dry grind process
Ethics, sustainable agriculture, 4*t*,
9*t*
Eucalyptus, pesticidal actions, 43*t*
Eugenol
pesticide activity, 43, 44*f*
toxicity, 45
Excitation resonance energy transfer,
mapping, to chlorophylls, 100*t*
Extracellular polymeric substances
(EPS), microorganisms, 54

F

Feed quality, distillers wet grain (DWG), 73, 75*t*, 76
Fermentations
  bacterial, 68
  simultaneous saccharification and, 67–68
  See also Dry grind process
Fluorescence emission
  low temperature spectra of envelope membranes, 119*f*
  phototransformation of Pchlide to Chlide, 123*f*
  spectra of ether extracts of pigments, 121*f*
Fluorescence excitation, ether extracts of pigments, 122*f*
Fluorescence resonance excitation energy, chlorophyll-thylakoid apoprotein assembly model, 98
Food Quality Protection Act (FQPA)
  pesticide management, 12
  pesticide tolerances in food, 42
Food wastes, maximum circulation, 11
Formosan subterranean termite (FST)
  determination of fungal pathogenicity, 176
  impact, 174
  location, 174
  microorganisms effective against, 174–175
  modification of fungal repellency, 185
  mortality of FST workers from strains of *Metarhizium anisopliae*, 186*f*
  Operation Full Stop, 174
  pathogenicity of FST workers, 178, 181*f*, 182, 183*f*
  recovery of microbes from termite cadavers, 182, 184*f*
  See also Entomopathogenic fungi
Fossil fuels, dependence, 16

Fungi
  atrazine-degrading, 136–137
  See also Entomopathogenic fungi; Green remediation

G

Gluconate-6-phosphate dehydrogenase
  activity in stroma, envelope, and thylakoid fractions, 114, 115*t*
  stromal marker enzyme, 111
Glycinoclepin A
  annulation method for synthesis of analogs, 169–170
  hatching stimulus, 163
  structure, 163*f*
  See also Soybean cyst nematode (SCN)
Green chemistry
  animal wastes, 11
  concerns for sustainable agriculture, 4*t*, 9*t*
  dependence of fossil fuels, 16
  desirable qualities for agriculture, 3–4
  environmental concerns, 15–16
  enzyme technology, 11–12
  food and urban wastes, 11
  future challenges, 13–17
  innovative approaches, 8
  maximum circulation of plant nutrients, 11
  natural product, 12–13
  need for increasing production, 14–15
  nutrient imbalances, 17–18
  outlook, 17–18
  population pressures, 14
  precision agriculture, 10
  principles, 8
  quality assessment system for agriculture, 18
  soil biological processes, 10
  soil degradation, 16–17
  sustainable agriculture, 6

subplastidic partitioning of
chlorophyll intermediates, 116*t*,
117*t*
Western blot, 114, 115*f*
See also Chlorophyll biosynthetic
intermediates
Thyme
pesticidal actions, 43*t*
toxicity, 45*f*
Thymol
pesticide activity, 43, 44*f*
toxicity, 45*f*
Toxicity
agricultural pests, 45
common plant essential oils to
cabbage looper, 45*f*
household pests, 45
mode of action for essential oils, 46
neurotoxicity, 46
Toxicology, adhesives, 61
Toxic waste streams, cleaning, 153–
154

U

Urban wastes, maximum circulation,
11
U.S. Environmental Protection
Agency (EPA)

commercialization of botanical
insecticides, 47
pesticide management, 12
reduced-risk pesticides, 42

V

Vegetation-enhanced bioremediation
long-term management, 129–130
See also Green remediation
Volatile organic compounds (VOCs),
adhesives, 53

W

Water
crop performance, 15
See also Atrazine; Green
remediation
Waterlogging, soil degradation, 16–17
Water primrose, atrazine uptake from
water, 131
Water resistance, adhesives, 55, 59, 61
Weeds, natural product chemistry, 5,
13
Wet milling, corn, 64